Howard W. Sams

Laser Design Toolkit

Dedicated to the memory of

Arthur Schawlow,

co-inventor of the laser,
who passed away as this book
was going to print.

Howard W. Sams

Laser Design Toolkit

By Carl Bergquist

PROMPT© Publications is an imprint of Howard W. Sams & Company, A Bell Atlantic Company, 2647 Waterfront Parkway, E. Dr., Indianapolis, IN 46214-2041.

International Standard Book Number: 0-7906-1183-X
Library of Congress Catalog Card Number: 98-68718

Acquisitions Editor: Loretta Yates
Editor: Pat Brady
Assistant Editor: J.B. Hall
Typesetting: Pat Brady
Proofreader: Stacy Nolan
Cover Design: Christy Pierce
Graphics Conversion: Terry Varvel, Kate Linder
Illustrations and Other Materials: Courtesy of the Author

PRINTED IN THE UNITED STATES OF AMERICA

9 8 7 6 5 4 3 2 1

Contents

INTRODUCTION

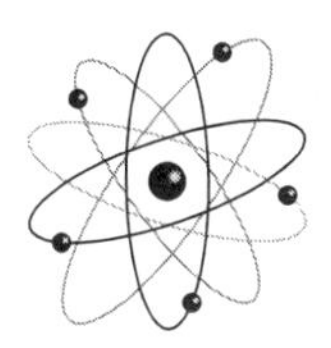

Ever since I was a “little shaver,” as some people would say, I have had a fascination with anything scientific. I mean, if it dealt with science, especially electronics, chemistry or physics, I loved it. Well, maybe not so much with physics. Some of that gets a little weird. But some of it is pretty neat, too.

Anyway, I have always been a fan of those fields. But when I first heard about a new device developed by Hughes Research Laboratories, the laser, that was it. This was something I could really get my teeth into. And I have never looked back! I'm just as enchanted with the laser today as I was back in 1960, and doubt that will ever change.

Now, you may be asking yourself, “Why? What, is this guy some kind of nut? I mean, in love with the laser?!” Well, I don't know about the “nut” part, but I do have some reasons, both pragmatic and sincere, to support my fondness for this device.

And I hope this text will reflect those reasons. Heck, I know it will, as there is so much to lasers it is hard not to at least “like” them once you get to know them. After all, they have vastly enriched our lives, even though we often don't recognize their contributions.

But that is only part of the story. For those who share my obsession with science, the laser can become a treasured hobby. There is an explicit thrill associated with watching that “red spot” on the wall, for the first time, as it emerges from a helium-neon laser you built. That is an event I can't begin to describe. You just have to experience it for yourself.

So in this book I introduce you to the wonders of the laser. To give you the chance to build and work with what has become an absorbing and remarkably inexpensive field. When I first became involved in constructing home lasers, the cost was substantial. But today it isn't any worse than many other areas of electronics.

Naturally, that makes this hobby abundantly more attractive. No longer do you have to take a second mortgage on the house to finance your laser lab. With the bargains available on the surplus and "previously owned" (Don't you just love that spin on "used?") markets, this venture can easily be yours.

For experimenters, one highly appreciated aspect of helium-neon lasers is that they run at nearly full performance for most of their anticipated life span. And, with most tubes, that is in the several thousand hour range. This makes buying used lasers an encouraging and reasonably safe prospect.

Naturally, used devices are less expensive than new ones, but you rarely sacrifice too much performance by investing in "previously owned" tubes. At least, not enough to be problematic concerning the projects covered here.

Speaking of those projects, they include building a helium-neon laser system, working with optics, constructing a light show, holography, building a laser light meter, working with laser diodes and more. All are guaranteed to entertain and, dare I say it, educate you. Of course, that is a "limited" guarantee. No, not really! Just a little humor!

I do unconditionally guarantee that if you are new to lasers, this book will really "trip your hammer." And, if you're an "old hand," you will like it anyway! OK, enough of the "hard sell." I sound like one of those TV infomercials!

Truly though, the world of lasers is a remarkable place to visit. If you have ever pondered the idea of venturing into this universe, I can only encourage you to charge forward without pause. I honestly doubt you will regret it. And, hopefully, this text will assist you in that journey.

I have such a captivation with these astonishing devices that every new excursion into the galaxy of collimated light becomes an adventure. And I sincerely hope that you too can discover that same euphoria and shoulder the same enthusiasm I've been so fortunate to experience.

Have Fun!
Carl

CHAPTER 1

LASER FUNDAMENTALS AND SAFETY

In this chapter, let's take a look at how lasers work, and how to safely use them. With all lasers, for that matter all light, the principle is the same, and we will discuss that in detail. And, by following some simple rules, many lasers are very safe to work with. That too will be discussed in detail.

Within our modern existence, the laser has become an inescapable factor. They are all around us, even though sometimes we aren't aware of their presence. Lasers do make life easier, but that is not what we are here to talk about. So, let's examine how these "enchanted light" devices do their stuff.

PRIMARY PROPERTIES OF LIGHT AND LASERS

Light, as we know it, is a collection of subatomic particles known as *photons*. These photons are produced whenever electrons in an atom change from an "excited" state to a normal state. Since they all start in the normal state, to get them in an excited state you have to excite them. Loudly yelling "BOO!" usually won't do the trick, so it is necessary to resort to some other method.

The best way I have found, to date, is to apply an excessive amount of energy to the atom which causes the electrons to "jump" orbit (normally from a lower to a higher orbit). To do this, the electron has to absorb some of the applied energy.

However, the electron doesn't like to be in the excited state, so it makes every effort to go home—home to its normal orbit or state. To do that, it has to yield the absorbed energy and, you guessed it, that energy is given off as photons. This is referred to as *stimulated emission of radiation* (SER).

In nature, the process is referred to as *spontaneous* emission of radiation, since no special efforts are made to achieve the electron orbital jumping. They are the result of whatever energy levels are present.

One of these mechanisms is responsible for every type of light we know, be it from an incandescent light bulb to the radiant effect of "dayglow" materials exposed to an ultraviolet source. And, it also accounts for the light a laser produces.

So, what makes a laser different from other sources of light? The answer to that question lies in a number of factors. First off, the illumination emanating from a standard light bulb is *omnidirectional*; that is, it spreads evenly in all directions (OK, OK! almost all directions). A laser, on the other hand, manifests a very tight beam that travels only in a defined direction.

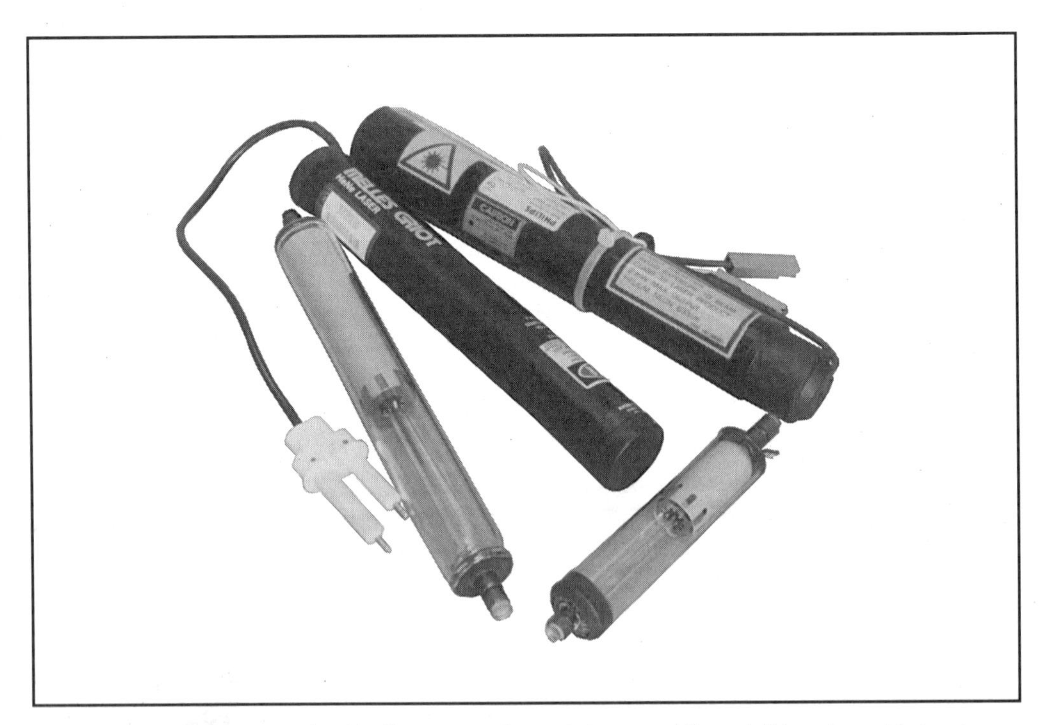

Photo 1-1. Some typical helium-neon laser tubes and "heads" (enclosed tubes). The tube on the left will be used in Chapter 2.

In order to accomplish this directivity, the light from a laser has to follow some rather explicit rules. In doing so, the device generates a very special type of light. Laser light!

The first requirement of laser light is that it be *monochromatic*; that is, light of a single color or wavelength. Most lasers do produce several different wavelengths, but one will prevail and overwhelm the others. We will talk more about this at a later time.

The second necessity is a physical principle known as *coherency*. Actually, laser light adheres to two separate types of coherency: spatial and temporal. The two work together to keep the emitted light together.

Spatial coherency can be thought of as the members of a marching band all staying in step as they march. With light waves, spatial coherency keeps the "crests" and "troughs" of the waves locked together. This, in effect, creates one single wave.

Temporal coherency refers to emission of these waves in very precise intervals, much in the same way an oscillator produces an output of a specific frequency. As a matter of fact, to laser technicians, the amplifying process is thought of as an oscillation function. Again, more will be said about this at a later time.

Finally, laser light has to be *collimated*. Collimation is, among other things, synonymous with *focus*. Think about people you talk to, that for reasons such as stupidity, mental defect or perhaps intoxication, don't make any sense. That would be normal light spreading all over the place.

In our case, we are talking with a person who keeps the conversation on track (does not diverge), and that results in good concentration. Aided by monochromisity and coherency, laser light does not spread like normal light. Thus, the beam is solid and well "focused" over a great distance. In other words, collimated!

When you combine these three optical "laws" of physics, the outcome is what we know as a laser. But this is not the entire story. There is another factor that not only contributes to the "LASER" acronym, but also is essential for the device to function correctly.

That factor is *light amplification* (LA). Just as your voice level can be increased by talking into a cone, or a small current can be changed into a larger current by a transistor, light too can be amplified. It just requires a little different approach.

This approach will be discussed when we get to the actual operation of a laser. But, for the moment, be aware that light amplification is necessary when producing true laser light.

IN THE BEGINNING

All right, where did all this theory come from? Well, electron orbit jumping has been known (suspected) for centuries and was regarded as spontaneous emission of radiation. Hence, laser theory evolved from that foundation.

But, to end up with a laser, some additional scientific genius was needed. This came from a paper written in 1916 by Albert Einstein (who else?). Good ol' Al understood all the light stuff we have pondered thus far, but that understanding left him with some very interesting questions about the performance of photons under certain circumstances.

For example, he had definite ideas regarding what would happen if a photon struck an atom in the "excited" state. He postulated that the atom would emit a second photon identical to the photon that hit it; thus, coherency and monochromisity would be achieved.

However, the abundance of naturally occurring "excited atoms" isn't that great, so he went on to propose the addition of an energy source to catalyze the process (said process is called "pumping"). Once enough excited atoms existed, collisions between the photons and these atoms would eventually result in a chain reaction of photon production.

Einstein coined this phenomenon *stimulated emission of radiation*, and it is the heart of a successful laser. It is also the "SER" component of the "LASER" acronym.

Furthermore, when a majority of the atoms present reach the excited state, a "population inversion" occurs. It is at this point that the laser's "threshold" is reached and the laser "lases."

AMPLIFY THAT LIGHT

One problem that accompanies stimulated emission is that collimation is not present. The light tends to scatter in all directions much as it does with a regular incandescent bulb. So, a method to concentrate the light was in order.

This is where the "LA" part of the name comes from; that is, *light amplification*. By confining the population inversion to a tunnel-like structure and placing mirrors at each end, the light will be reflected back and forth between the mirrors. Some photons will still escape from the sides, but the majority will maintain their "back and forth" trek.

This concentrates the excited atoms and elevates the chances of photons colliding with them, thus producing more photons. Since a small number of photons are creating a larger number of photons, the definition of "amplification" is met.

Now we have this tunnel structure with all these photons bouncing back and forth and striking more atoms and producing more photons, and that is all well and good. But since the photons are trapped inside the structure, no light can escape. Not a good deal if you're trying to produce a laser.

However, if some of the reflective surface is removed from one of the mirrors, this will furnish an escape route for some of the light. The amplification will continue, as many of the photons will be reflected back, but the fleeing particles depart the device as a laser beam.

In laser terminology, this mirrored structure is known as the *resonance cavity,* and that term gives rise to the previously mentioned *oscillation*. Virtually all laser types employ resonance cavities, except some special gas systems and the very high-powered devices which produce such an abundance of excited atoms that mirrors are not necessary.

APPLICATION OF LASER THEORY

In theory, a brick can be made to "lase" if enough energy is applied. You may well destroy the material, but it will lase, at least for an instant. An example of this is the "x-ray" lasers developed as weapons at the Lawrence Livermore National Laboratory.

Here, a strip of thin metallic foil, acting as the *laser medium*, is bombarded with super-high energy from either a huge neodymium glass laser or an underground atomic explosion. The energy coming from either source vaporizes the metallic foil, producing an intense x-ray beam. The medium is destroyed in the process, but the laser beam is intense enough to decimate anything in its path.

However, with most lasers, the medium is not destroyed, or even damaged. Hence, with proper operation, they can perform for many years, often with very little, if any, maintenance.

Modern lasers are seen in four distinct varieties: Solid-State, Liquid, Gaseous and Semiconductor. While all are important, the gaseous and semiconductor species lead the pack in frequency of commercial use.

The *solid-state* devices usually employ rods made from either synthetic ruby or neodymium YAG (yttrium aluminum garnet) crystal or neodymium "doped" glass as the laser medium. These are pumped with light. Intense flashes of white light are applied by surrounding the rod with xenon "flash tubes" similar to, but more powerful than, the tubes employed in photographic strobe units. These systems can produce high-energy "pulses," or in the case of the glass rod units, "continuous beams" capable of burning holes in many substances. Needless to say, they can be dangerous and need to be handled with appropriate caution.

Liquid lasers are almost entirely confined to the "dye laser" category which utilizes fluorescent dyes as a medium. These systems are highly versatile, as they can be "tuned" to wavelengths from infrared to ultraviolet. But, degradation of the dyes and system complexity do tarnish that versatility.

Since the dye lasers are high in maintenance and expensive to build and operate, they have seen little demand in the commercial sector. Both these and the solid-state lasers are primarily used by the scientific community for research purposes.

Gaseous lasers provide, by far, the widest variety. They come in such flavors as helium-neon, carbon dioxide, copper vapor, argon, nitrogen, neon, krypton, carbon monoxide, noble gas, helium cadmium, iodine, xenon-helium as well as some nasty fluorine (F) based systems usually called "chemical" and/or "excimer" lasers. And these are just a few of the total number of gaseous lasers.

Each one has its own purpose and value, but several stand out as the most important species of this class—at least commercially speaking. Those would be the helium-neon, carbon dioxide, argon and nitrogen lasers.

We will get to the first candidate in a moment, but let me say a few words about the other three. Carbon dioxide (CO_2) systems are used for large-scale metal cutting and welding, as they can produce beams in the kilowatt range. They also produce infrared wavelengths which can't be seen by the human eye, so visible light "aiming" lasers are usually included in the package. CO_2 lasers are dangerous, and must be used with extreme caution.

Argon (Ar) lasers produce light in the blue-green visible range, and can produce beams with hundreds of watts of power. There is an old axiom about looking directly into the beam of an argon laser. You will only do it twice; once with each eye. Yes, it's true! These babies will blind you in a hurry, and should be considered nothing short of downright dangerous.

Argon lasers have found many practical uses in fields such as medical research and fingerprint identification. But, they do require very careful handling and suitable eye protection.

Nitrogen (N) lasers are the oldest of the ultraviolet wavelength devices, dating back to the early 1960s. They are often employed as "pump" lasers for dye systems and are so simple to build they have been featured in scientific magazines.

They are also one example of a laser that doesn't absolutely require a twin-mirror resonance cavity. However, mirrors do increase their power output to a few hundred milliwattts maximum. Eye safety is not as critical with nitrogen lasers as it is with CO_2 and Ar systems because the lens of the eye tends to absorb and filter out ultraviolet light. Nonetheless, a prudent rule to follow is NEVER LOOK DIRECTLY INTO ANY LASER!

Our last gaseous laser, the *helium-neon* (HeNe), is by and far the most widely employed variety of the category. They have found extensive application in universal product code (UPC) readers, such as seen at checkout counters, and optical reading storage devices like CD/CD-ROM players.

The first helium-neon laser appeared in 1961, just months after the first working "ruby rod" laser was demonstrated. This device functioned at 1.15 nanometers (near infrared), but scientists soon followed up with the more traditional red 632.8 nanometer tubes.

Although they are feeling competition from the semiconductor lasers, HeNe systems are still in wide use. This makes them ideal for experimenters, as the tubes and power supplies are readily available and CHEAP!...well, relatively cheap, anyway.

Since we are going to build just such a system in Chapter 2, let's dissect the helium-neon laser to find out how the basic principles apply. This exercise should help you better understand the workings of lasers in general.

First, take a gander at *Figure 1-1*, the anatomy of a helium-neon laser tube. As can be seen, the assembly consists of an elongated glass envelope that contains the necessary parts and gas. The large metal tube is called the "spider" and it acts as both the cathode of the tube (is electrically connected to the cathode terminal post) and a support for the bore.

The center glass tube is the *bore*, and it has a very small internal diameter. It is this diameter that largely determines the size of the emitted beam; thus it is kept to less than a millimeter. One end of the bore terminates in the metal anode connection post, while the other end is left open inside the spider which allows the gas mixture to enter.

If all this sounds a lot like a diode, congratulations! The HeNe laser tube, like a number of gaseous lasers, is basically a specialized diode.

On each end of the hollow terminal posts are the mirrors. One, of course, is fully reflective, while the other is only partially reflective. With most of the tubes I have worked with in the last ten years or so, the beam emits from the anode end, but many texts describe it as departing from the cathode. Each manufacturer has its own style, and it doesn't take a rocket scientist to figure out which end is emitting the beam. I mean, just look at the ends! But, not directly into them! Remember, SAFETY FIRST!

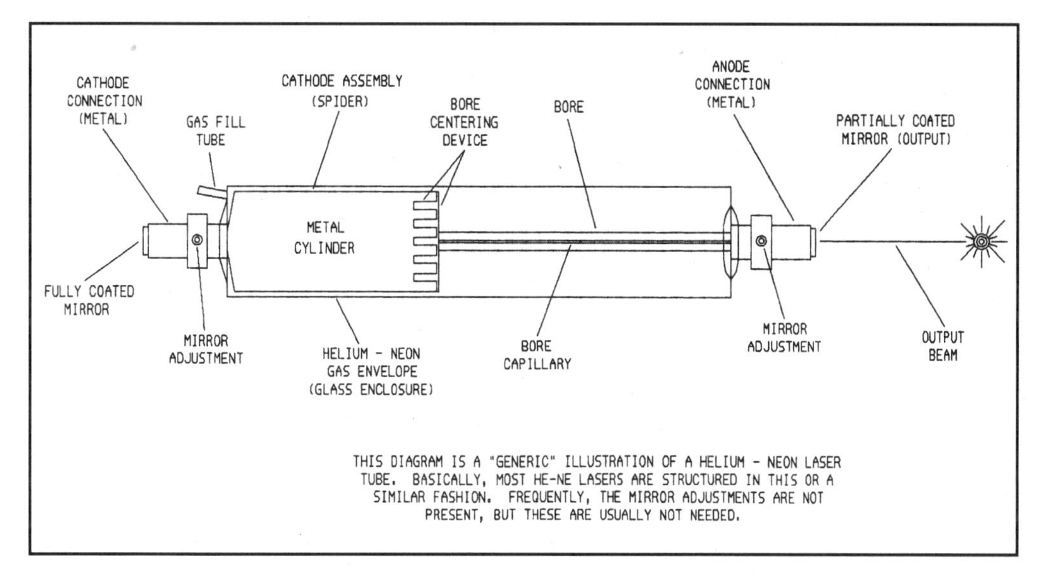

Figure 1-1. Anatomy of a helium-neon laser tube.

This being a "universal" picture of HeNe laser tube structure, I have also illustrated mirror adjustment collars. These are not often seen on small tubes (less than 10 milliwatts) these days, but many of the older tubes did have them. And, some of the higher-power tubes still have them. However, each tube is aligned at the factory for maximum performance, and it is best not to monkey with these adjustments unless the laser doesn't seem to be producing the amount of light anticipated.

Lastly, the envelope is filled (through the gas fill tube) with a combination of helium and neon gas. A generic ratio of about 10 to 1, respectively, is considered normal. I say *generic*, as this can vary widely among assorted manufacturers.

Additionally, the gases are routinely placed in the tube under a slight pressure (less than an atmosphere) to facilitate better operation.

With that under your belt, let's see how we get this sucker to lase. First, a high-voltage direct current (DC) power supply is needed. It must produce between 1 and 4 to 5 kilovolts at about 3.5 to 7 milliamps (depending on the size of the tube). It must also be able to generate "short-term" voltage pulses in the 3 to 10 kilovolt range. These are used to start the tube and are referred to as the *kicker* or *trigger* voltage.

A word of caution here! THIS IS A NASTY POWER SOURCE! Nasty in the sense that while it probably won't kill you, it is definitely going to talk to you. The amperage is low, but let me say from personal experience that tangling with the business end of a HeNe laser power supply, well....it ain't no day at the beach!

Hell! It hurts! And, it can do damage to sensitive and/or delicate body tissue. So, be careful when working with these supplies. More about this in the safety section.

OK, with a suitable supply's negative lead connected to the laser's cathode terminal post, and positive lead to the anode post, apply power and the laser should light up. It might take a second or two, so be patient.

This is due to the fact that the high voltage is energizing the gases to an excited state, which in turn produces a population inversion. You know, those little electrons are jumping orbits all over the place. Sort of like rats jumping a sinking ship.

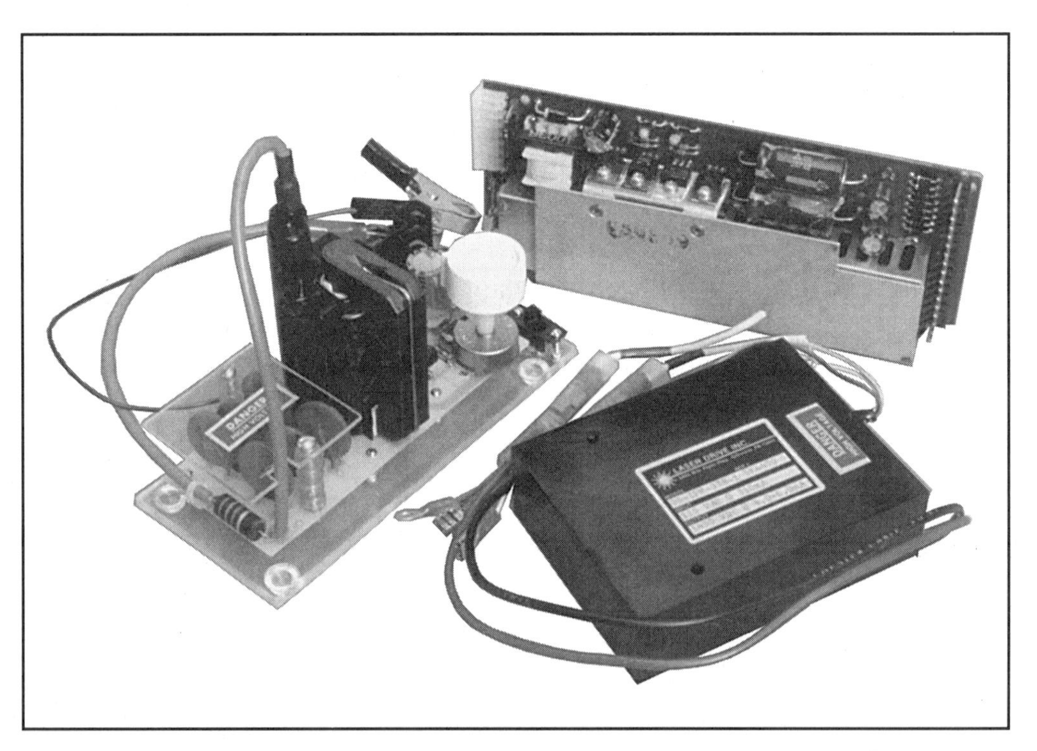

Photo 1-2. Some typical helium-neon laser high-voltage power supplies. The one on the left is the kit in Chapter 3, while the one in front is used in Chapter 2.

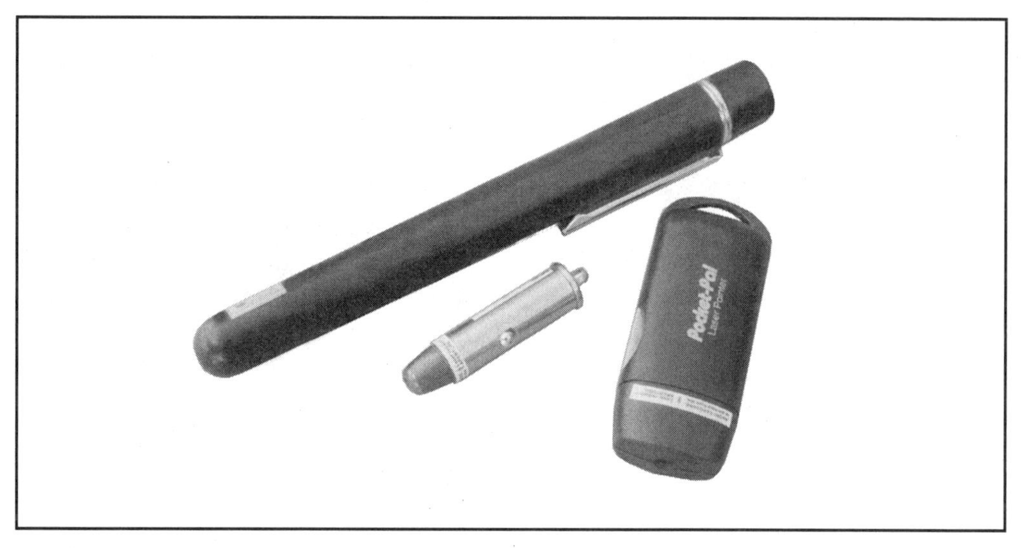

Photo 1-3. Three different "laser pointer" styles. These use laser diodes, and as can be seen, can be quite small. Each is a self-contained diode laser system.

Anyway, at one end there should be a thin very intense beam emerging. That is because some of the photons are escaping the system through the partially reflective mirror. And, since they are coherent, monochromatic and collimated, they will appear on a nearby wall as a tiny spot of bright red light (a wavelength of 632.8 nanometers to be exact).

As you look into the tube (through the sides), you will notice the bore is glowing a dazzling pinkish-yellowish-orange. This is the "lasing" action. Since most HeNe lasers have about 1 percent efficiency, if you could measure the intensity of the glowing bore against the beam, the bore would be about 100 times brighter.

That efficiency is surely nothing to write home about, but it is acceptable because of the unique characteristics of laser light. Since lasers can perform so many important feats with their amazing light, it is only fair that we cut them a little slack when it comes to efficiency. I mean, it's not as though we are trying to light up the room with them.

And, that as they say...is that. I don't know who "they" are, but there must be someone out there somewhere saying stuff like that. If you know who, please write me!

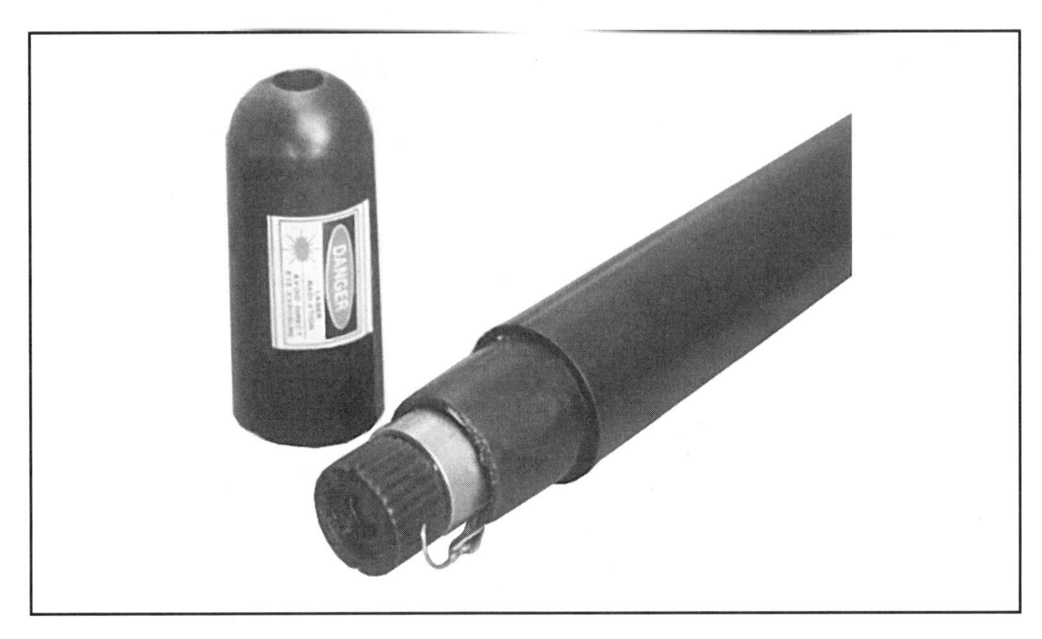

Photo 1-4. A closeup view of the diode module inside one of the laser pointers in *Photo 1-3.* The module contains the diode, driver and collimator lens.

Our second "most employed" laser type is the semiconductor, or diode lasers. Actually, these are by no means second. As an educated guess, I would venture to say that diode lasers outnumber all others combined by about 20 to 1. (All right, be nice about the educated part! I'm almost positive I made it through at least the sixth grade. Ha, ha, ha. Just a little humor to break the tension.) Their modest cost, low power requirements and size have made laser diodes very attractive to both hobbyists and manufacturers. They are common today in laser disk players, CD-ROM drives, CD players, fiber-optic communication and laser printers.

Interestingly enough, semiconductor lasers date back to 1962 with research at the Bell Laboratories and in the Soviet Union. The first ones, like the first LEDs, emitted in the infrared region. However, for many years certain obstacles involving wavelength and operating temperature prevented them from becoming the familiar item we know today.

Describing the structure of a semiconductor laser is difficult as there are so many distinct technologies and configurations. Their origin is found in the field of light-emitting diodes (LEDs), and most are a hybrid of the gallium arsenide (GaAs) junction that made LEDs possible. Usually, trace amounts of phosphorous (P), aluminum (Al) or indium (In), or combinations of all three, are added to produce the desired emissions.

There is also a group of specialized diodes called the *lead-salt diode* lasers, but these are not exactly hobby items. They lase in what is known as the "far infrared" region (3,300 to 29,000 nanometers) and have to be operated in cryogenic temperatures. That, combined with their expense, makes them devices we do not want to mess with.

Back to standard laser diodes. About as "general" as this gets is that these diodes are small blocks of semiconductor material(s) less than a millimeter square (see *Figure 1-2*). The older models emit their beam from the edges, while some newer designs emit from the surfaces.

As with all lasers, stimulated emission of radiation and a population inversion are both necessary for proper laser operation. Hence, small mirrors are either etched

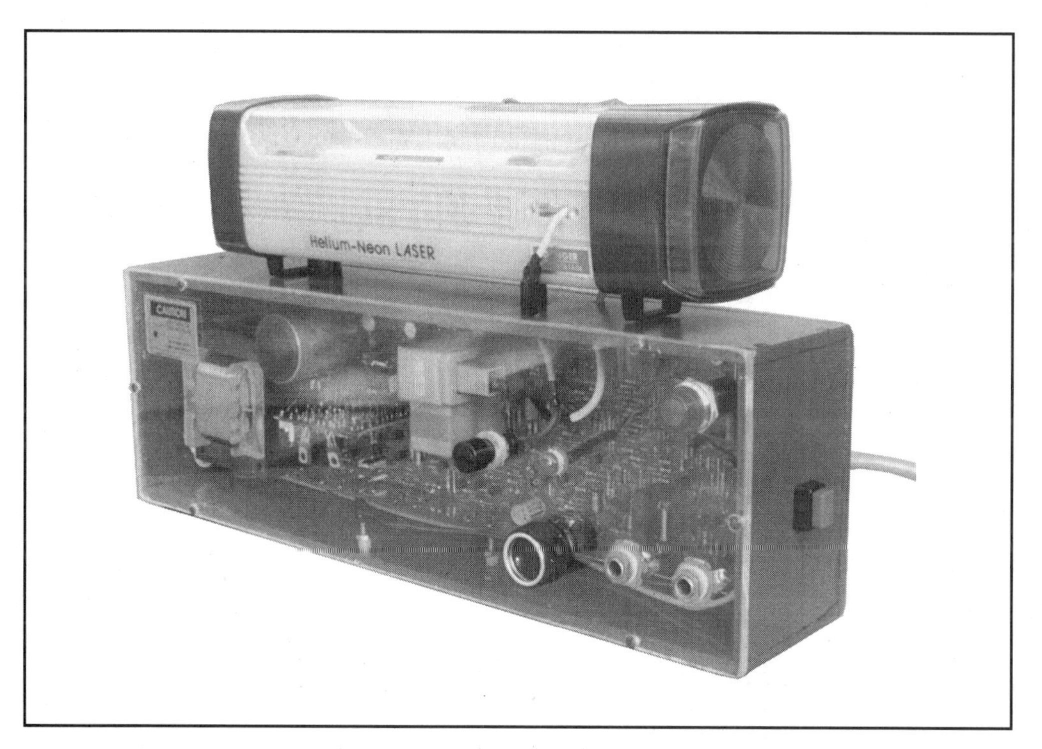

Photo 1-5. The Siemens HeNe lasing system. Rather elaborate, this device will produce between 2 and 4 milliwatts of collimated red light.

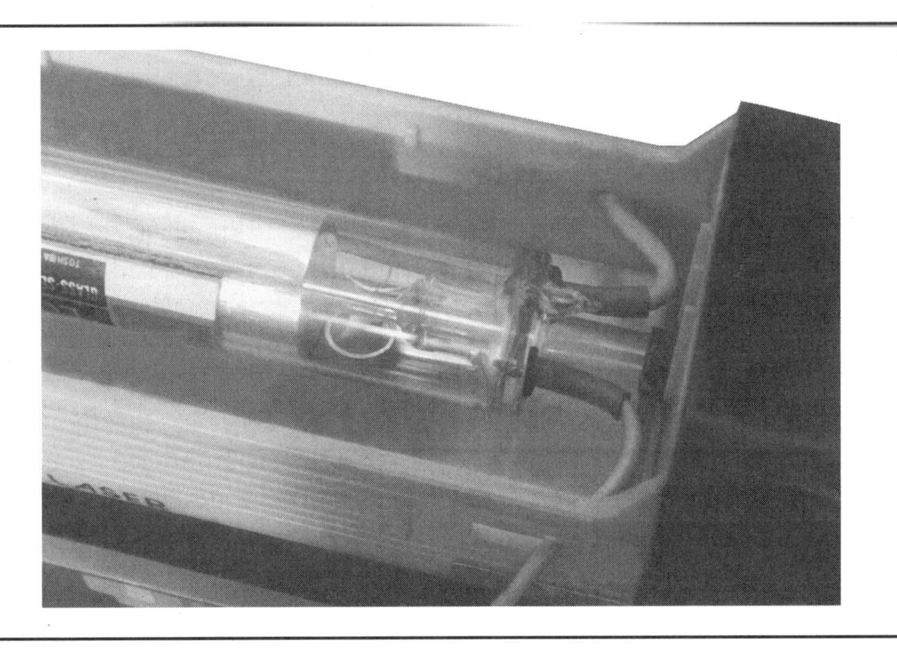

Photo 1-6. Closeup of the Siemens HeNe tube lasing. Note the glow is confined to the center core, or bore, of the laser.

or cleaved into the structure. This furnishes reflection back into the junction cavity which promotes light amplification and produces laser light.

Applying very carefully controlled currents to the assembly results in a "forward biasing" of the diode. The consequence is the release of energy in the form of photons—unlike silicon diodes that release the energy as heat. The photons are then bounced back and forth between the *mirrors*, and the beam escapes through the partially reflective surface.

Most commercial diode lasers also incorporate a photodiode in their structure. This is done to prevent excess current damage to the delicate junction. By monitoring the light intensity and sending those results to a "feedback" network, the photodiode can control the current flow. This not only protects the laser, but also keeps the light output uniform.

As with light-emitting diodes, the wavelength is determined by the composition and structure of the junction. To date, most visible light laser diodes are in the red wavelengths (670, 660, 650, 635 nanometers); however, yellow (594 nm) and

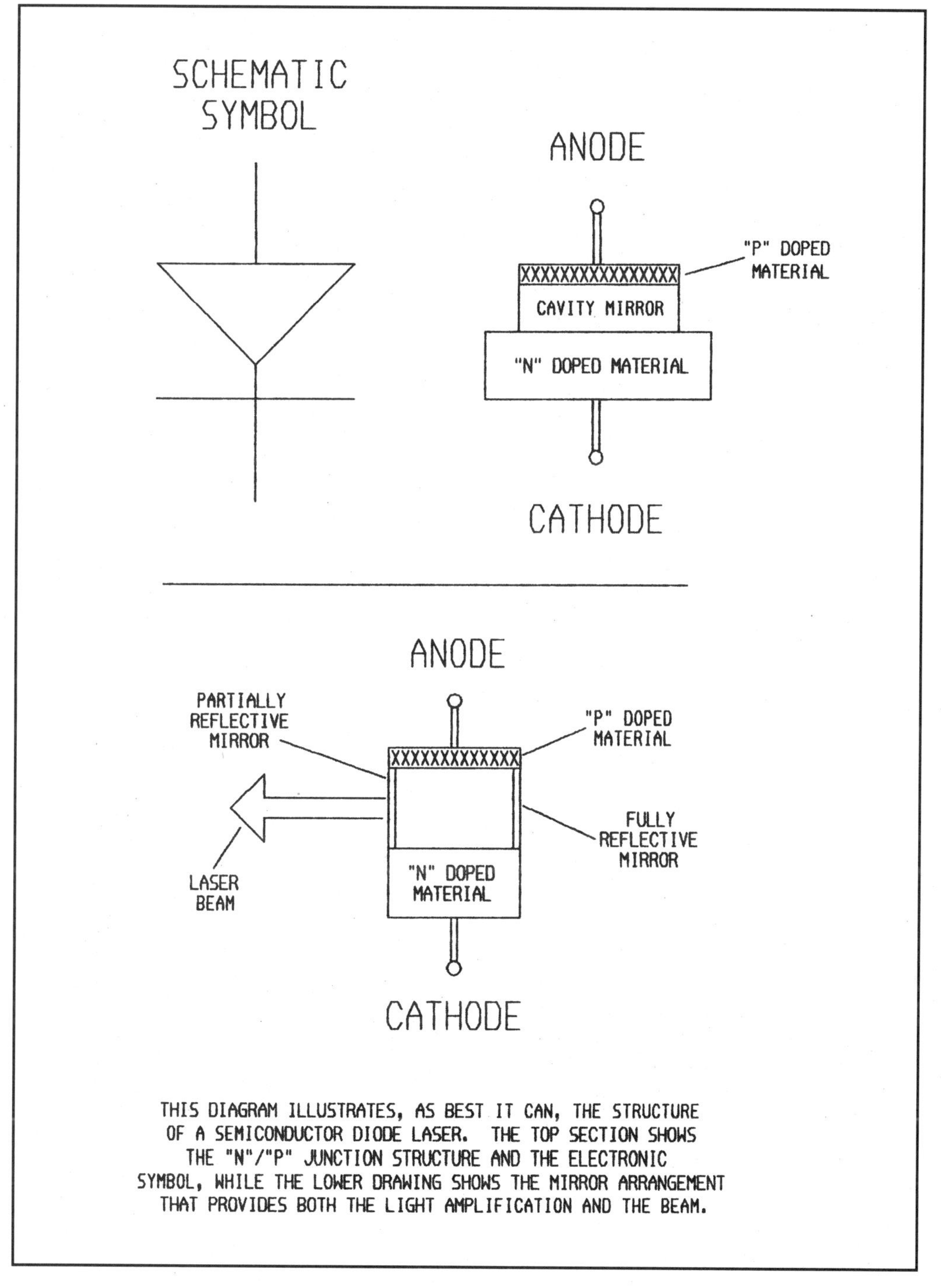

Figure 1-2. Semiconductor diode laser structure and electrical connections.

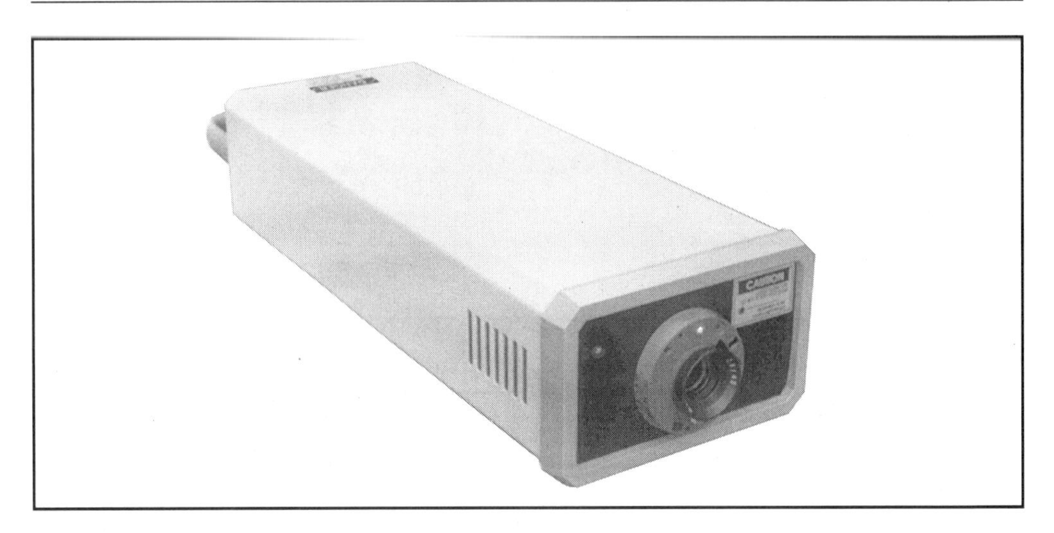

Photo 1-7. A "laboratory" style laser system. It produces around 1 milliwatt of power and is highly portable.

green (532 nm) are available for special needs. Yellow is used for reading disks in the new digital versatile drive (DVD) ROM drives, and green is seen in "super" bright laser pointers.

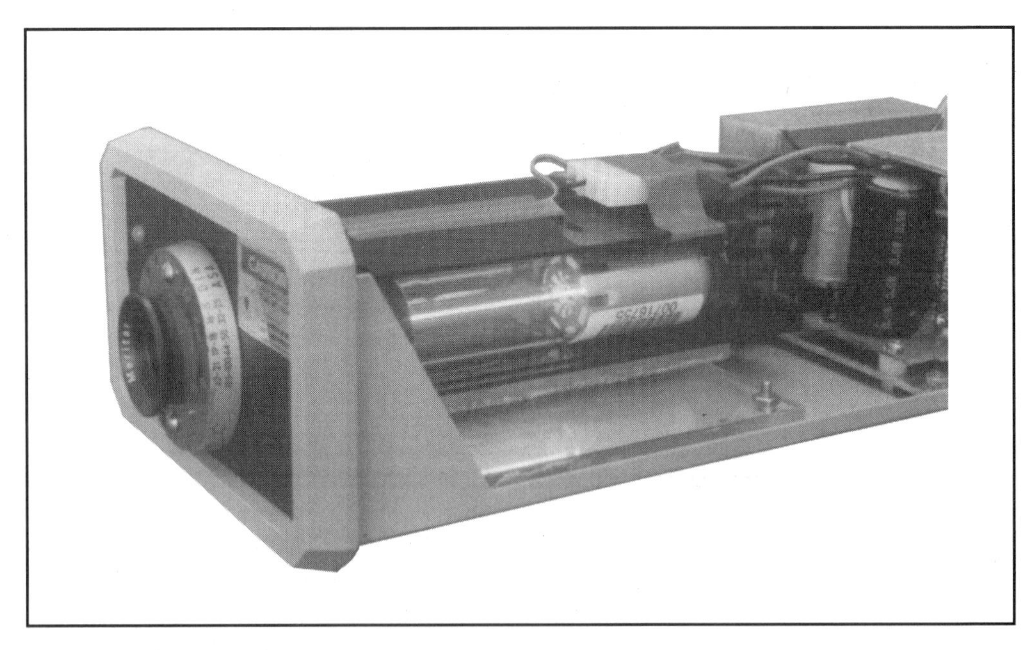

Photo 1-8. An interior view of the laboratory laser system. The 12V DC power supply is on the right and the high-voltage PS/laser tube assembly is on the left. Note the glowing laser bore and support "spider."

Of course, the infrared region is also widely represented with numerous individual wavelengths. Many of these are employed in standard CD/CD-ROM drives, telecommunications, printers and security systems. These are, of course, invisible to the human eye (except for some 780 nm units that produce a dull red glow).

Through the use of "quantum well" structure, some very high-power diode lasers are coming on the market. We are talking about 1/2 to 1 watt infrared units. These powers are the result of the extremely tiny quantum well junctions that may only be a few atoms wide. At such a minuscule size, energy transfer is exceptionally efficient, and these junctions can produce substantially elevated power output levels.

Laser diodes generally come in two types: *gain guided* and *index guided.* Restricting current flow to a narrow strip of the laser medium constitutes gain guided operation, while index guided diodes confine the light through refractive-index changes. Each has its place, but index guiding is the most commonly employed technique today.

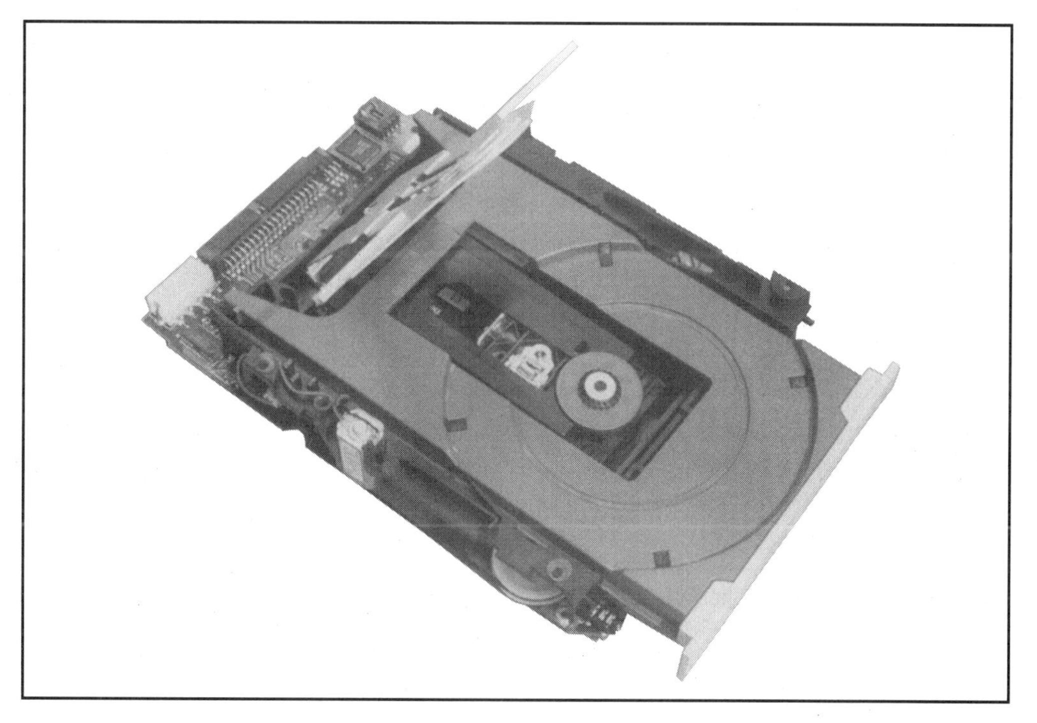

Photo 1-9. A typical CD-ROM drive. Note the laser assembly in the center of the device (white in color).

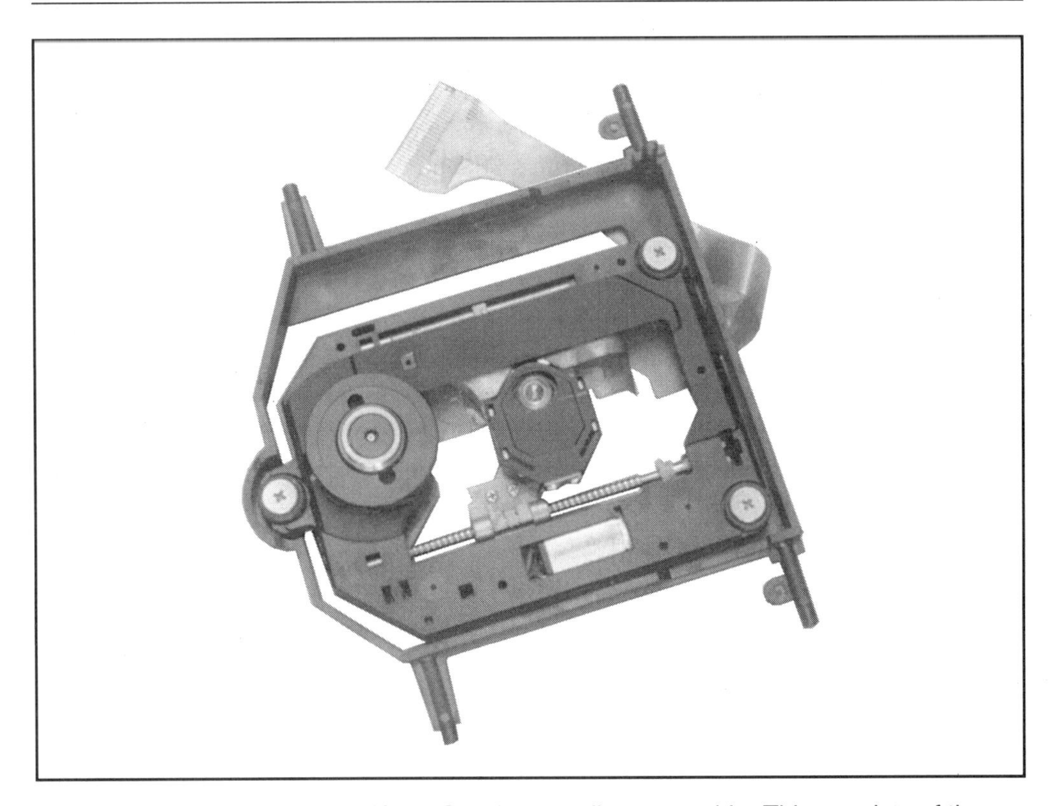

Photo 1-10. A typical CD/CD-ROM data reading assembly. This consists of the stepper motor (left) and the laser/lens unit (center).

When it comes to the structure of the junction itself, three primary approaches have emerged over the years. The first was the *homojunction* diode, where the junction is made from a single composition (GaAs, for example). This design was only viable at cryogenic temperatures, and then was so inefficient, it soon fell by the wayside.

The next attempt was the *single-heterojunction* laser. This utilized a "sandwiching" arrangement employing materials of different chemical composition. Some still exist, but they can not operate continuously and are seen mostly in military equipment.

The last and latest approach is the *double-heterojunction* diode, in which all active layers are sandwiched between layers of different chemical composition. This strategy allows for continuous operation at room temperature and is the principal structure used for most modern diode lasers.

At the risk of stating the obvious, we will continue to hear more about laser diodes. Research continues into smaller, more efficient units with a better selection of wavelengths. I won't be surprised to see orange, blue and white laser diodes in the next few years, just as we have recently seen blue and white LEDs added to the existing inventory of red, orange, yellow, lime, green, aqua and magenta.

SAFETY, SAFETY, SAFETY!

While I have left this inordinately important subject for last, perhaps I should have started this chapter with safety. However, it does help to first understand the workings and operation of laser systems to fully appreciate this topic.

Lasers are a fascinating subject, with many interesting and important applications for both the hobbyist and professional. And, as a hobby, they will prove most rewarding. The roads that lead out of this megalopolis can take you on some mighty absorbing journeys, and that is what hobby electronics is all about—at least for me, anyway.

However, I would be negligent in my responsibilities to you as my readers if I didn't stress the importance of safety measures. I would hate to see you lose interest in, or be scared off from, this remarkable subject because I didn't present the proper caveat regarding the dangers of lasers and laser light.

On the other hand (examining the flip side of this coin), I don't want to effect the same result by informing you of these hazards. They are present, but if you adhere to the simple precautions outlined herein, there is nothing to fear from the lasers you construct from this text.

I am mainly concerned that you comprehend the dangers, which I have already touched on, before you work with laser equipment. Basically, there are two areas of possible trouble: *shock* from high voltage and damage to the *retina* of your eye. So let's go back over each domain individually.

In terms of the laser projects in this book, electrical shock is only a concern with the helium-neon system. As stated earlier, a HeNe power supply has to produce a minimum of around 1,000 volts for the smallest tubes, to as much as 5,000 volts for the large 50 to 75 milliwatt tubes. There is also the 3 to 10 kV trigger pulse.

While the amperage is low (on average from 3.5 to 7.5 milliamps), any of these supplies will give you a nasty shock if you touch the leads. Additionally, most supplies have fairly large capacitors that store energy, and the HeNe tubes themselves retain the high voltage even after the power is removed.

So, after you shut down a system that is not enclosed in a case, it is best to short the laser tube's cathode to anode. This can be done with an alligator clip-style jumper lead and will discharge both the laser tube and the power supply capacitors. Connect one clip to the cathode post and the other clip to the anode terminal. But, do this with caution as there may well be some sparking when you first touch the lead to the laser's anode. Also, leave this jumper in place for several seconds to completely discharge the system.

For enclosed lasers, and/or laser "heads," there isn't much you can do aside from letting the unit discharge in the normal fashion. Many power supplies employ resistors to assist in this process, but it does take time. So, the best advice is not to fool with anything inside a closed system for at least 30 minutes after power down.

Even at that, if you do open a closed laser, it is always best to manually discharge all large capacitors and the tube before doing any serious work on the system. And that applies to a unit that hasn't been operated for a while. Some of the capacitors used in these supplies are capable of holding large enough charges to *ZAP* you for remarkable periods of time. As I stated earlier, arguing with a "hot" high-voltage laser supply is a sucker's bet. It's also an argument, and a bet, you are destined to lose.

Anyway, in review, the essential messages here are: 1) be extra careful not to touch, or even get near, the leads of a functioning high voltage laser power supply, and; 2) after shutdown, be sure the supply and tube are completely discharged before you tinker with any of those pretty components, or the tube. Follow those rules closely and you won't get eaten by the evil monster that lives in the cave down by the sea.

One final note concerning laser power supplies. If you eventually get involved with such things as argon and carbon dioxide lasers, take extra care with their supplies. Argon lasers, for example, run on voltages in and around 300 volts, but the amperage can be in excess of 60 amps. And that, my friends, can/will kill you dead!

The second "health" concern lasers pose is possible vision damage. Due to collimation, all the energy is concentrated in that threadlike beam the device produces. Much like focusing the sun's rays to a pinpoint with a magnifying glass to set paper and wood on fire, that highly condensed beam will easily enter the eye and burn the retina. This damage is often irrevocable and can lead to blindness.

Hence, as has been previously stated, DO NOT LOOK DIRECTLY INTO THE BEAM OF ANY WORKING LASER!!! I strongly recommend that you live by this advice, even though many will tell you it is safe to look into helium-neon beams of less than 1 milliwatt. Don't take the chance! Your vision is too important to lose as the result of carelessness!

With that said, let me talk a little about the degrees of danger regarding various wavelengths, power ratings and laser designs. Generally speaking, visible light is the most dangerous to the eye. In the infrared and especially in the ultraviolet ranges, the lens and other body tissue tend to filter, absorb or neutralize the beam.

For example, blood within the human body is an excellent filter for infrared (and red) wavelengths, while short-wave ultraviolet can penetrate only certain materials (quartz and sodium chloride for example). Thus, the lens of the eye blocks such light.

On the other hand, just about all visible wavelengths easily invade the eye and can do damage. Remember, if you can see the light, it is getting to your retina. This doesn't mean that a quick glance down the beam of a 1 milliwatt helium-neon laser is going to blind you for life. It probably won't. But, being a red laser, it is more likely to do harm than an ultraviolet device. Again, though, it's best to heed the previous advice and avoid looking into laser beams.

The laser's power is the next consideration. Staying with the helium-neon example, while a 1 milliwatt beam will probably be fairly harmless to your eye, a 5 milliwatt system can and will damage the retina if the exposure is long enough. Exactly how long "long enough" is depends on a number of factors. These include the systemic condition of the individual eye, the angle at which the beam is observed and the type of beam scrutinized.

Since the eye is most sensitive to the green and blue wavelengths, these are the ones to most arduously avoid. I mentioned before that you will only look directly into an argon laser twice—once with each eye—and the wavelength is one of the reasons. Of course, the 200 milliwatt to 20 watt (or higher) beam ratings also have a lot to do with it.

Naturally, there is a government agency that sets standards and evaluates laser equipment. Why not? Another way to spend our tax dollars! This one, the C.D.R.H., which stands for *Center for Device and Radiological Health*, probably does a better job than most. At least, it seems that way. Also, the Food and Drug Administration is mixed up in this some way or another, but I have been unable to get a straight answer as to exactly how.

The center sets ratings for laser emissions, and *Table 1-1* details those ratings. Also, you must place an appropriate label on all lasers as per the C.D.R.H. parameters, but more will be said about this in the laser construction chapters.

As for the diode lasers, most of these are rated at 3 milliwatts or higher, so they need to be treated with respect. This is especially true for the diodes that have a collimating lens assembly installed. The "raw" laser diodes do not collimate light as well as other lasers, so they are less dangerous. But once the optical elements are in place, the beam is tightened up and becomes more hazardous.

Once again, don't look into the little window on top of these diode modules when they are on. Even without the collimator, these gems could harm your retina(s).

Incidentally, laser pointers are sometimes touted as the ultimate cat toy. It is true that cats (especially kittens) love to chase the red spot, but their eyes are just as sensitive as ours. In fact, they are even more sensitive, so if you plan to use a laser pointer to amuse you cat, please observe the same safety precautions you would for yourself. You wouldn't want to blind Kitty, or Sylvester, or Fluffy, et cetera.

And that pretty much covers the dangers of lasers and their light—at least the ones we will be working with. As I said, the rules are simple to follow, and doing so will help protect you against the hazards they present.

LASER CLASS	POWER OUTPUT
II	≤ 1 MW
IIIA	≤ 5 MW
IIIB	≤ 1 W
IV	VERY DANGEROUS PULSE/CW SYSTEMS
V	≥ 1 KW

THIS TABLE ILLUSTRATES THE BASIC RATING CATEGORIES AS SET BY THE C.D.R.H. THIS COVERS MOST OF THE COMMONLY AVAILABLE COMMERCIAL AND SCIENTIFIC LASER SYSTEMS. SOME OF THE SPECIAL PURPOSE AND/OR MILITARY LASERS MIGHT NOT BE COVERED.

Table 1-1. Center for Device and Radiological Health (C.D.R.H.) laser power ratings.

CONCLUSION

I hope the preceding information has been of value to you. Lasers have come a long way since that day in 1960 when Theodore Maiman demonstrated his ruby rod device. Since then, scientists and engineers have explored a vast cosmos of laser possibilities, often producing highly pragmatic equipment.

Today, lasers are used to heal the sick, improve telephone communications and record and read vast amounts of valuable information. They are also used for entertainment in the form of light shows, CD players and laser disk systems. And then there are the other uses, such as keeping the foundation and walls of that new house you are building straight, or keeping the perimeter of your property safe from intruders.

These, of course, are only a few of the multitude of ways the laser has benefitted society. A little quiet thought will quickly bring to mind many other applications. But, more to our purposes, and the point of this text, lasers are the foundation of a fascinating hobby. I presume that since you are reading this book, you either know that, or you have a burning desire to learn more about lasers.

In either case, BRAVO! First, I don't care what they say (There's "they" again!), but knowledge is not always bad. And second, I doubt you will be disappointed with the projects in this text. Whether you are new at this or an "old hand," lasers are forever a source of amazement, enchantment and just plain fun for those who are willing to give them a try.

CHAPTER 2

A HELIUM-NEON (HeNe) LASER SYSTEM

I can't think of a better introduction to the world of lasers than the helium-neon (HeNe) gas variety. These have been around since the early 1960s and are familiar to most of us as that red "threadlike" beam wandering around inside the universal product code (UPC) readers you see at checkout counters.

Literally millions of these tubes have been manufactured to accommodate the vast number of UPC readers in service, and in that lies one of the reasons this gem is just right for laser experimentation; namely, there are so many of them around that they are CHEAP. At least, cheap by laser standards.

You can see what gets my attention when it comes to hobby electronics! Aside from the price though, helium-neon tubes are highly practical for numerous laser projects. In addition to superbly demonstrating laser principles they also display every attribute and property of this amazing type of light.

Hence, the subject of this chapter is a fully operational helium-neon laser. You will be able to construct this system without breaking the bank, or robbing it, and the finished product provides the basis for many of the text's remaining projects.

Another amenity involving this project is safety! We will be constructing a 1 milliwatt (mW) rated unit, and according to major HeNe tube manufacturers, these are "eye safe." That is to say, the beam from tubes of this strength will not have an adverse effect on your eye's retina. However, as previously mentioned, I would not advise staring directly into the beam regardless of what the manufacturers claim. They could be wrong, you know!

Secondly, the high-voltage power supplies for lasers of this intensity do not generate enough "juice" to kill you. Well, at least they won't kill most people. I can't say for certain they are safe for everyone, but while the voltage is high, the amperage is so low that a shock from the supply shouldn't be fatal.

Again, as previously mentioned, stay away from the tube, leads and high-voltage section of any laser supply while it is functional. It may not kill you, but it is going to inflict substantial pain if you tangle with that potential. Trust me, I speak from experience which was the result of sheer carelessness.

So, let's get to the heart of this chapter and build a "for real" laser. If you have never before done this, you are in for a pleasant surprise, as there is an inescapable thrill about seeing that red dot on the wall for the first time—especially when it comes from a laser YOU have constructed.

THEORY

Since I covered the scientific theory of a gas laser in Chapter 1 (you remember, physics 103), I will avoid boring you with a repetition of that. Instead, let's examine the hardware we will need for our project.

First, we want to lay our hands on a 1 mW laser tube. Both the surplus market and laser specialty companies (see source list) carry extensive lines of HeNe tubes. Some are used (equipment "pulls"), but many can be purchased brand spanking new. In either case, the cost is remarkably low (often less than $25) and most suppliers test the used units to be sure they are functioning properly. For the prototype, I went with a used Spectra-Physics model 088 (see *Photo 2-1*).

Incidentally, there is nothing wrong, bad, disgraceful, stupid or unwise about using formerly-owned tubes (Notice I employed a more polite term for "used.")! I have bought many a "recycled" tube and haven't been accused of, well... most of that stuff. At least, not so far, anyway.

If you get the laser from a reputable source, it should operate commendably for many years. The average HeNe tube has a life expectancy of between 15,000 and 20,000 hours and most will not show unacceptable performance until very late in their lives.

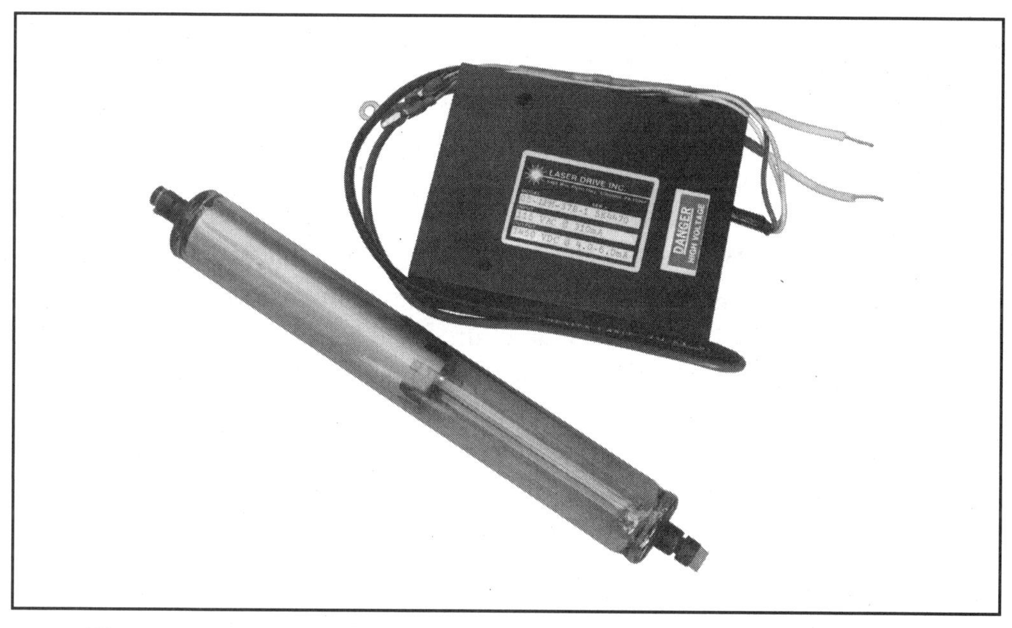

Photo 2-1. The two primary components of the helium-neon laser system: the laser tube (left) and the high-voltage power supply (right).

However, many companies carry new tubes for about the same price or slightly higher, so if the thought of buying a "not new" unit bothers you, get a new one. That way, you will also get a full warranty and perhaps some peace of mind.

Next, we will need a high-voltage (HV) supply to activate the tube. Again, the surplus/specialty market is brimming with said supplies. And, again, they can be purchased either new or used (I chose a surplus Laser Drive Inc. model 05-LPM-378-1). For my third "again," if you buy used, the *good* companies will be sure you get a working unit. However, should you encounter trouble, they will help you with the problem and replace the supply if necessary.

Now, here's some good news! Chapter 3 (you know, the next one) will show you how to build your very own high-voltage supply from a kit. If you would like to go this route, combining two chapters into one, this may well be your ticket. What'a ya think? Huh...huh! What'a ya think? Huh? (Just excuse me. I do this from time to time. It's nothing serious!)

By the way, for some more good news, I have only included "good" companies in the source list. These are all suppliers I have personally done business with, and to date, none of them have taken me for a ride.

OK, back to the power supply. A 1 milliwatt tube is going to require between 1,000 and 1,300 volts DC at 3 to 3.5 milliamps, with a *starter* voltage of between 6,000 and 8,000 volts. So, you need to keep this in mind when considering a supply. Virtually all the epoxy-encased (potted) units will run this size laser, but some are capable of running higher-power lasers as well.

Now, powering a 1 mW tube from an oversized supply will work. I mean it won't hurt anything. But price goes up proportionally regarding supply capability, thus it isn't wise to "overkill" the HV source, unless, of course, you're planning on moving up to a higher-power laser at some time in the future.

Anyway, depending on the type of HV supply you acquire, you may also need a low-voltage power supply. Many of the laser supplies are designed to work off standard household voltage (120 VAC) but some do need a 9 VDC, 12 VDC or other separate direct current source.

Upon locating a HV supply that appears to meet your needs, be sure to confirm its input voltage requirements. Many give you a tolerance of, say, 9 to 14 VDC, and for 1 mW tubes, a standard linear regulated supply will easily do the trick. Check Chapter 3 (*Figure 3-2*) for the schematic of just such a supply. That one delivers 12 volts at a maximum of 1.5 amps.

Now that you have the individual components of this system, it's time to put them together. So, the next stop on this "magical mystery tour" is the construction phase. On that note, let's get right to it!

CONSTRUCTION

This part is pretty easy. I mean, basically, these gems all go together in the same fashion. *Figure 2-1* illustrates the process, starting with the initial power source (either 120 VAC or a DC unit) going to the high-voltage power supply, which in turn goes to the helium-neon tube.

Note that a *ballast resistor* is required between the positive (anode) high-voltage terminal and the anode of the laser. This resistor is used to "trim" the HV output, ensuring operational stability, and to prevent damage to the tube due to voltage and/or current excesses.

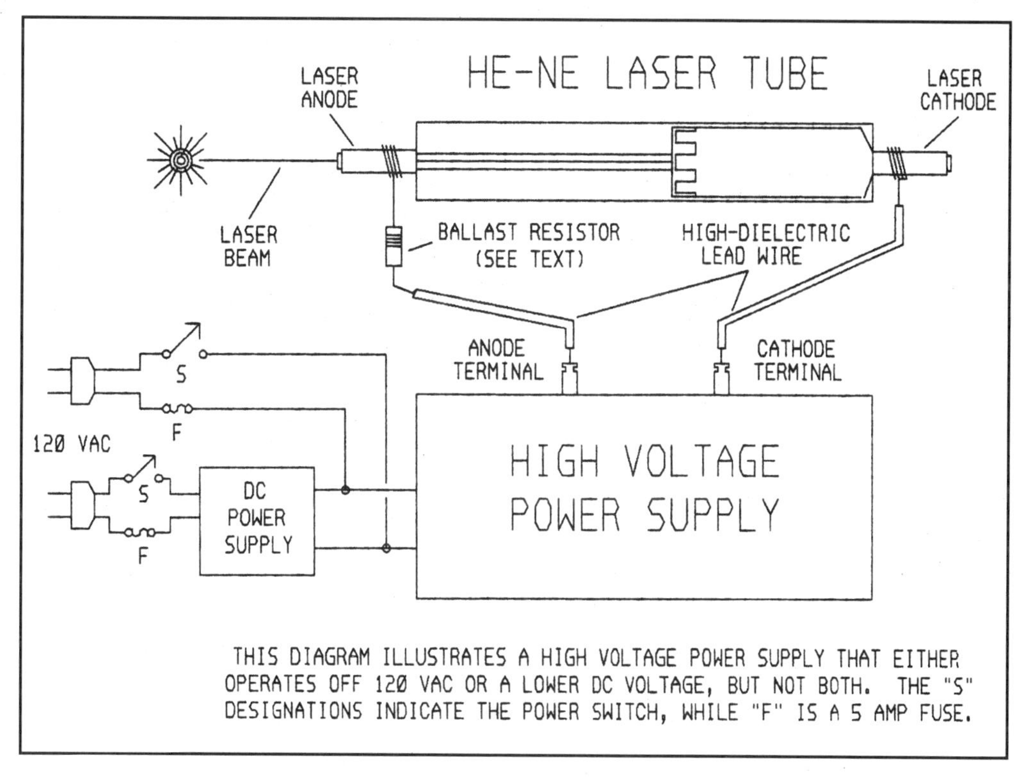

Figure 2-1. HeNe laser system.

Ballast resistors are usually rated at 2 to 5 watts in the 60 to 110 kilohm range, and the term refers to supplying *balance* or *equilibrium* to the high-voltage section. If you go the *laser head* route—that is, a tube sealed in a protective enclosure—the ballast resistor is almost always included inside said enclosure. This is because the resistor should be positioned as close to the tube's anode terminal as possible.

As for our project, a 2 watt ballast resistor of about 75,000 ohms should fill the bill. If the tube wants to "sputter" (flicker) or seems to run exceptionally hot, you may need to respectively decrease or increase that value. A little experimentation will reveal the proper resistor for your tube.

The cathode side of the HV supply is simply connected to the laser's cathode terminal. Be sure to use *high-dielectric* (high-voltage) leads for both connections. This type of wire can be secured from some of the laser suppliers or many local electrical and/or electronics dealers.

HELIUM-NEON LASER PARTS LIST

LASER TUBE

Any of the many 1 milliwatt helium-neon tubes available on either the surplus or laser specialty market. The source list contains a comprehensive register of reliable companies that handle either new or used tubes. Some of the major manufacturers include:

Spectra Physics
Hughes
Siemens
Melles Griot
Uniphase

HIGH-VOLTAGE POWER SUPPLY

For a 1 milliwatt helium-neon laser, you will need a high-voltage supply that provides around 1,000 to 1,300 VDC as the continuous operating voltage. Additionally, the supply must be able to produce a momentary starting or "kicker" voltage of between 6,000 and 8,000 VDC. The amperage rating will need to be about 3 to 3.5 milliamps. This is, of course, the standard for a 1 milliwatt laser, and higher-power lasers will require higher ratings. Again, the source list will furnish many companies that handle both new and used supplies.

POSSIBLE LOW-VOLTAGE POWER SUPPLY

Some of the high-voltage supplies require a low-voltage DC input, such as 9, 12 or 24 volts. This possible requirement must be considered when buying a high-voltage supply. Chapter 3's *Figure 3-2* illustrates a highly suitable regulated supply for a 1 milliwatt helium-neon laser.

OTHER COMPONENTS

A 2A fuse, fuse holder, 16 gauge line cord, sheet plexiglass (1/16 inch and 1/4 inch thicknesses), clear PVC pipe, clear plastic enclosures, high-dielectric (high-voltage) cable, SPST power switch, hardware, etc., are some of the other "stuff" you will probably need for this project.

One last caution! DO NOT try to solder the high-voltage wires to the tube's terminals! Since these terminals are usually heat fused to the laser's glass envelope, heat from a soldering iron is very likely to crack the glass. This will allow the gases to escape (gas out) and the tube will be kaput!

While not absolutely, positively, unequivocally, utterly or unconditionally needed, a 2 amp fuse (F) is a good idea. If something goes wrong, this will protect the system, especially the power supply(s), from a current surge. Also, an ON-OFF switch (S) is another good idea. However, I don't think I need to go into great detail as to why. At least, I hope not!

As for the safety factor! It is best to house your tube in a suitable protective enclosure such as a length of clear PVC pipe. This will not only shield the tube from accidental breakage, but also keep you, or someone else, from touching it when the laser is on. And, if the pipe is transparent, the bore's *lasing* action will still be visible.

Regarding the high-voltage supply, if it is of the *potted* version (encapsuled in epoxy), then it is "shock safe" as long as you stay clear of the anode/cathode leads. If you use a surplus "open" style supply, or the unit described in Chapter 3, some precautions should be observed.

For example, I strongly recommend an appropriate case to isolate the supply from individuals observing the system (like some idiot who reaches in and grabs hold of the thing). Again, if you employ a transparent housing, the idiot, uh...*individual*, observing the system will be able to see the power supply.

Check out the supply in Chapter 3, and you'll note that "nonconductive" plexiglass is used to guard the high-voltage section, and the completed unit is mounted on a plexiglass base. Even though these measures do help prevent possible shock, a clear housing would assuredly be a sound notion for this supply.

The only other consideration concerns the power cord. It should be something in the 16 gauge range as opposed to the lighter more common 18 gauge cord. Also, it never hurts to use a 3-conductor cable with a ground line. This allows trouble to be routed to ground instead of through the system, or worse, YOU!

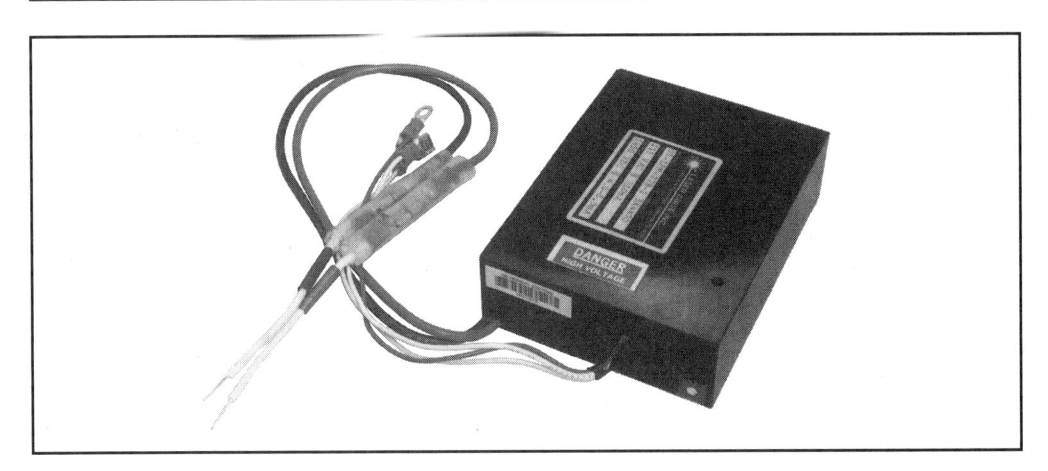

Photo 2-2. The Laser Drive, Inc. "potted" high-voltage power supply used for the helium-neon laser system.

The last step in the construction process is to place a Center for Device and Radiological Health (C.D.R.H.) warning label on the laser. *Figure 2-2* illustrates a number of examples for this label which can be photocopied if desired. Basically, the *type*, *power*, *class* and *wavelength* need to be included, as well as a *direct-exposure* admonition. If you like, you can make your own labels following those parameters.

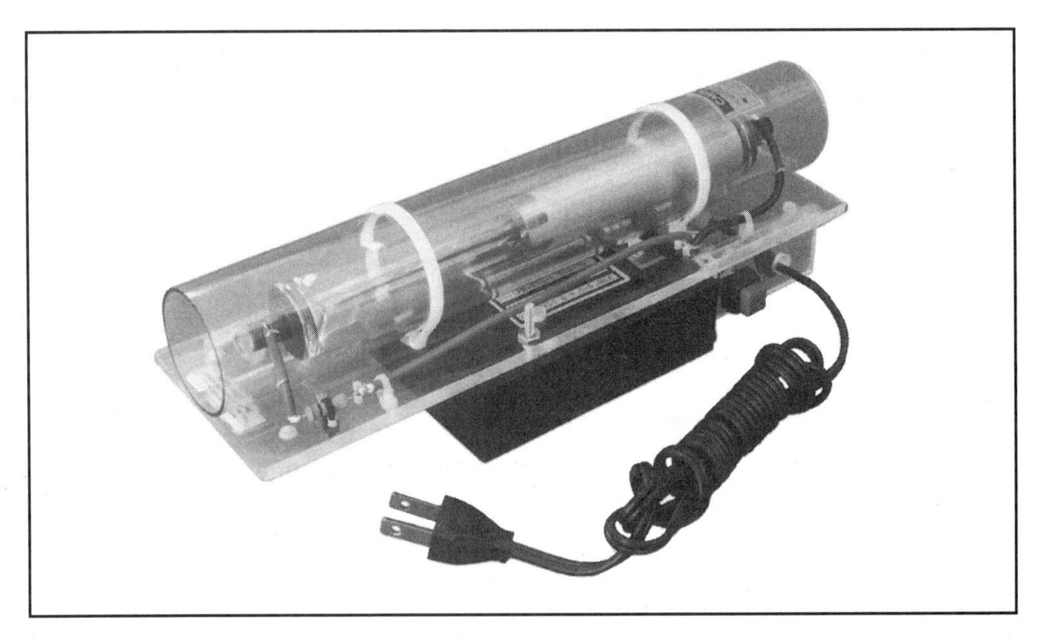

Photo 2-3. The completed helium-neon laser. This system includes the tube, high-voltage power supply and ballast resistor mounted to a clear plexiglass base.

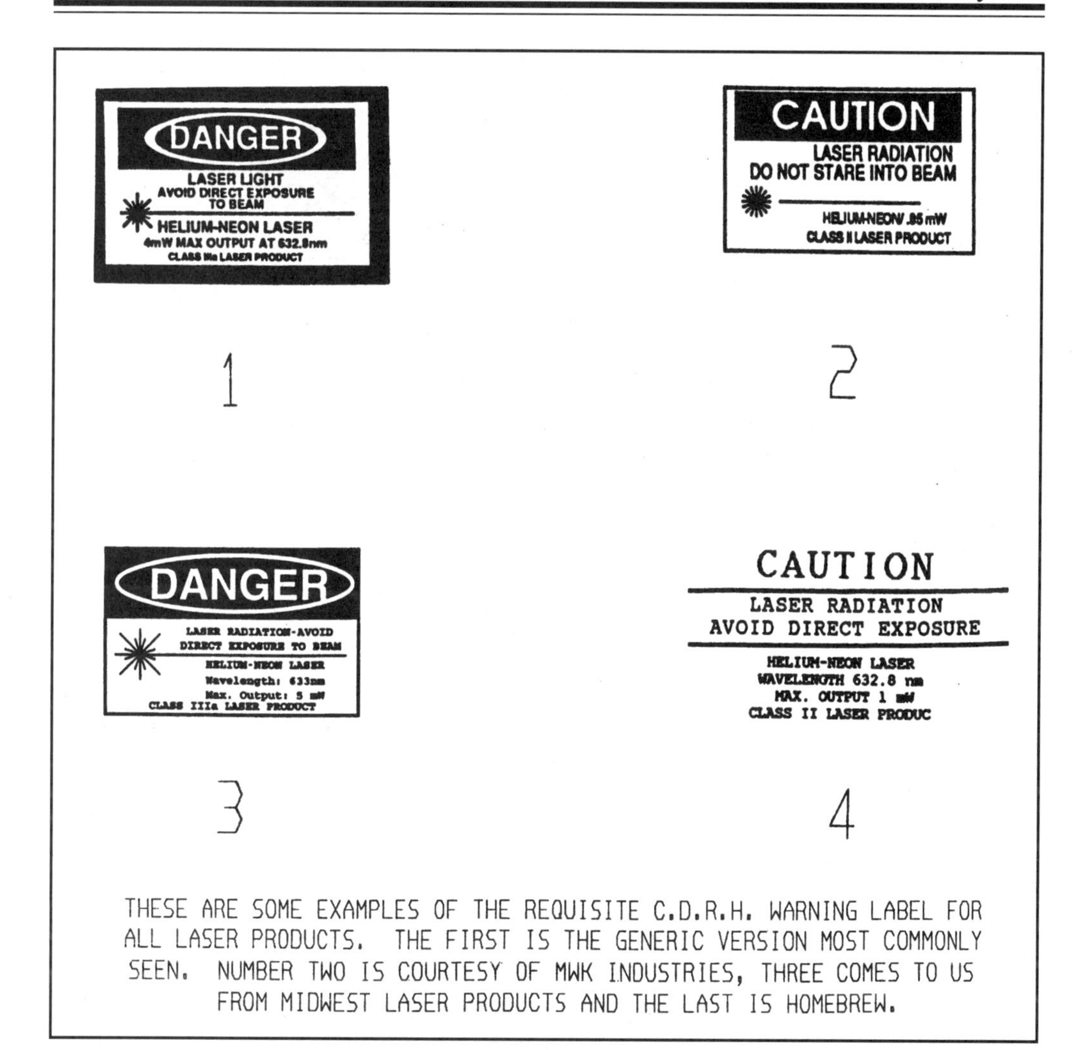

Figure 2-2. C.D.R.H. warning labels.

TESTING AND OPERATION

Now that you have everything hooked up, plug the cord in and hit the power switch. If all is well, the tube's bore will glow brightly, and the laser will project that notorious "red dot" on whatever it is pointed at.

Presuming all that happens, your system is functioning correctly. If your supply has a current adjustment, like the unit in Chapter 3, and the tube wants to "sputter" (flicker on and off), adjust the output current potentiometer. That usually solves the problem.

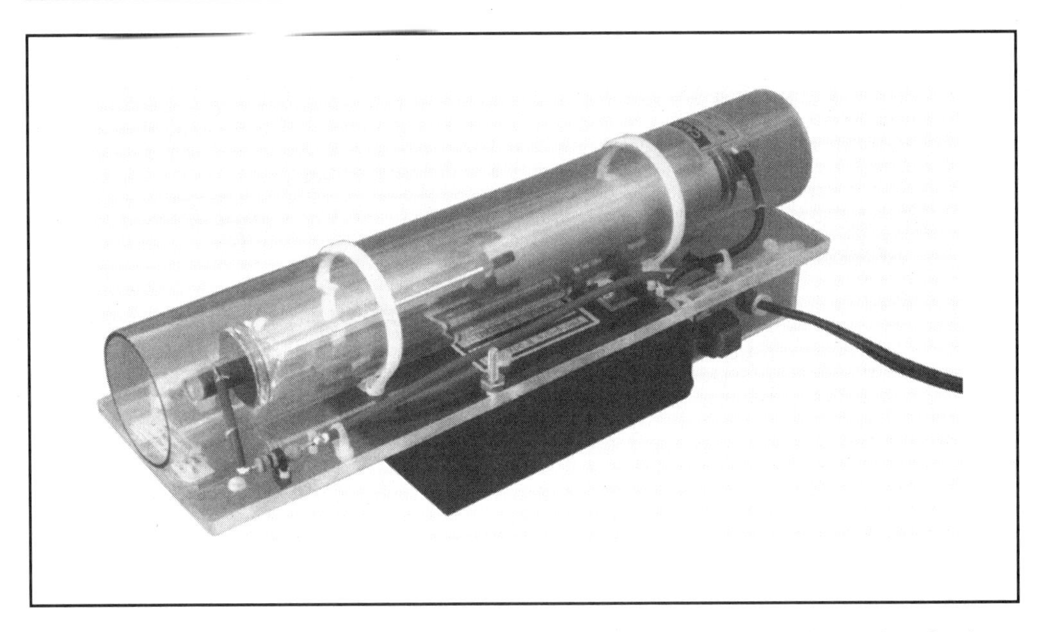

Photo 2-4. The completed helium-neon laser system in operation. Note the glowing "bore" of the tube.

You should also be able to detect "speckling" around the dot. This is primarily due to the fact that the beam is striking an uneven surface. It may not appear uneven to you, but laser light attains such a precise resolution that the symmetry of anything it encounters is clearly illustrated by this effect.

Speckling, of course, manifests itself as a "glittering" pattern that surrounds the dot. It will vary in intensity depending on the type, shade and reflective quality of the target material, and as has been stated before, is a reliable test of whether the beam is indeed true laser light.

With that pretty red dot on the wall, you have a working helium-neon laser. And, it is high time for the two of you to get better acquainted. Try reflecting the beam off different objects; however, be careful with strongly reflective surfaces like windows and mirrors. You don't want the beam coming back into your eyes.

As you proceed through this exercise, take note of the way laser light reacts to the various materials. This information will be of enormous use in future experiments and projects. Also, notice the amount of absorption demonstrated by individual surfaces. Again, data of this type will be important later.

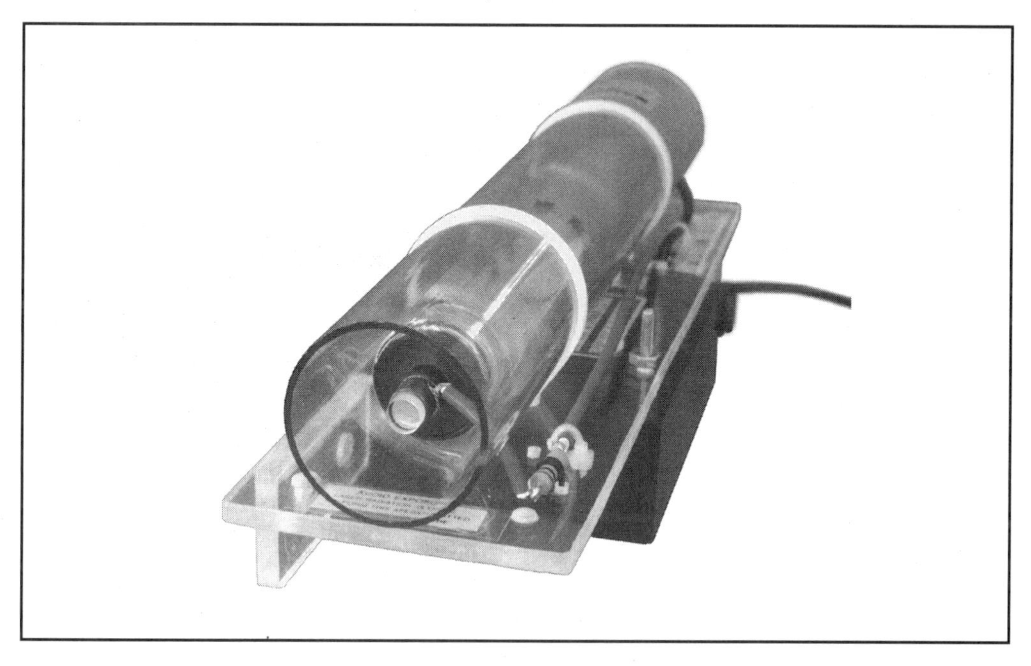

Photo 2-5. Another, more "down the barrel," look at the completed helium-neon laser system in operation.

Now, try casting the beam on a more distant object, like a nearby building. This maneuver is best done at night, as the dark provides enhanced visibility. Daylight, especially bright sunlight, tends to "wash out" the red dot, making it harder to see. One caution here! Avoid shining the beam through windows where it might strike someone.

With most 1 milliwatt lasers, you should be able to "hit" structures up to a mile away. And that includes "low flying" clouds. Also, if you can enlist the aid of an assistant, ask him/her to measure the size of the dot on that distant object. This will give you an idea of the laser's *beam divergence*; that is, the amount the beam spreads over distance.

On the average, 1 milliwatt beams will spread between 1 and 3 feet for a 1 mile span. This naturally will vary among various available tubes. But for the most part, that amount of divergence can be expected. If it is considerably more, there is a problem, probably with the mirrors, and the tube should be returned to its source for replacement.

CONCLUSION

Alright, you are on your way towards many hours, days, perhaps even years of fun and frolic with your newly constructed helium-neon laser. Chapter 4 will familiarize you with the basics, and many of the remaining chapters address some of the numerous fascinating experiments you can perform with your new "toy."

Laser light is unrivaled in terms of purity and collimation and in that lies a spectacular ability to perform so many of the feats for which it is renowned. And, fortunately for all of us, HeNe lasers admirably demonstrate those properties. I say fortunately because they are both available and economical.

Additionally, with the proper precautions, they are safe to operate. That, I think it goes without saying (but I'll say it anyway), is immensely important for anyone interested in learning about and/or experimenting with lasers. With helium-neons, you're not constantly worrying about beam hazard, as you are with argon or ruby pulse lasers. This is not to say, however, that appropriate precautions should be cast to the wind! Not at all. Keep safety incessantly in mind, even with "safe" lasers.

Forgive me for constantly dwelling on safety, but it is a crucial element of the equation. Your encounters with this field will be far more rewarding, and informative, if you don't hurt yourself. At least, that's usually the way it works for me!

In any event, with this system in hand, you will be able to journey into a truly mystical world—a world involving light, distinctly different from all other types. Light that holds cryptic secrets awaiting your discovery. With this system in hand, those discoveries will unfold like hidden treasure for your ultimate astonishment. And, all because of the properties of this extraordinary form of illumination.

CHAPTER 3

UNIVERSAL HeNe LASER POWER SUPPLY KIT

OK guys, here is the kit for this text. As you saw in the last chapter, in order to function, those interesting looking helium-neon laser tubes do require a source of high-voltage/low-amperage power. We used an epoxy-encased "block" supply the last time, but this project allows you to actually construct the supply. And, that helps you better comprehend how these systems operate.

This kit comes to us from Marlin P. Jones & Associates, is simple to build and reasonably priced. Most important, though, is that it, well....works! I'm not going to name names, but over the years I have encountered numerous problematic laser power supply kits. Often they do not perform as advertised, or worse, they do not perform at all.

When that happens, it's not only frustrating, but it can make you downright angry, livid, irate, furious, incensed, enraged and *expletive-deleted.* Hence, I looked for a kit that wouldn't do any of that to you. And here it is. This is not to say that other kits will not perform as well—just that I can personally vouch for this one.

As a quick sidebar, I have also tried a myriad of "start-from-scratch" circuits, usually found in magazines and textbooks. While some did work quite well, they usually employed very specialized and/or hard-to-acquire components, especially when it came to the high-voltage step-up transformers. Since that is always a "pain in the posterior" when constructing equipment, we are much better off with a reliable kit.

This unit is just that, and will power lasers up to about 3 milliwatts. That should cover most of the tubes you will be working with, at least at this point. So, with that said, let's take a look at how the power supply does its job.

THEORY

Examining *Figure 3-1*, the schematic diagram, will reveal a fairly generic system; that is, an oscillator is used to simulate an alternating current (AC) voltage, which is boosted by a transformer (TV "flyback" variety). That voltage is then stabilized, and sustained, by the resistor/capacitor *ladder network* (C5-10/R5-10).

The oscillator is based on what should be an old friend; the LM555 oscillator/timer integrated circuit (IC). The frequency, or *duty cycle*, is controlled by resistors R1, R2, capacitor C1 and potentiometer R11. The output from pin 3 is sent to transistor Q1, through resistor R3, which amplifies the current to the level the transformer demands.

Adjustment of R11 allows you to change the oscillator's duty cycle, which ultimately changes the output of the supply. This is done through a process called *pulse-width modulation*, where the duration of the individual pulses control the amount of power delivered. This provides adjustment of the supply to comply with a variety of tube requirements.

Once the potential has been transformed (by T1) to a sufficient level, capacitors C5 through C10 and resistors R5 through R10 serve to maintain a consistent output. That is important with laser supplies, as uniform operation is essential when delving into territory such as holography and spectroscopy.

As a last step, the final output is trimmed with ballast resistor R4. As you learned in the previous chapter, the term *ballast* is applied here in the same way it would be with ships; that is, to provide stability or balance.

So, this circuit is fairly straightforward. Nothing in the way of revolutionary technology is involved. Just a highly dependable method of producing the requisite voltage and current needed by helium-neon laser tubes.

CONSTRUCTION

The kit comes with everything you will need, including a "tinned" printed-circuit board (PCB), all parts and lead wires. The first step, though, is to take a component inventory to be sure everything is there. The kit I received had it all, but you

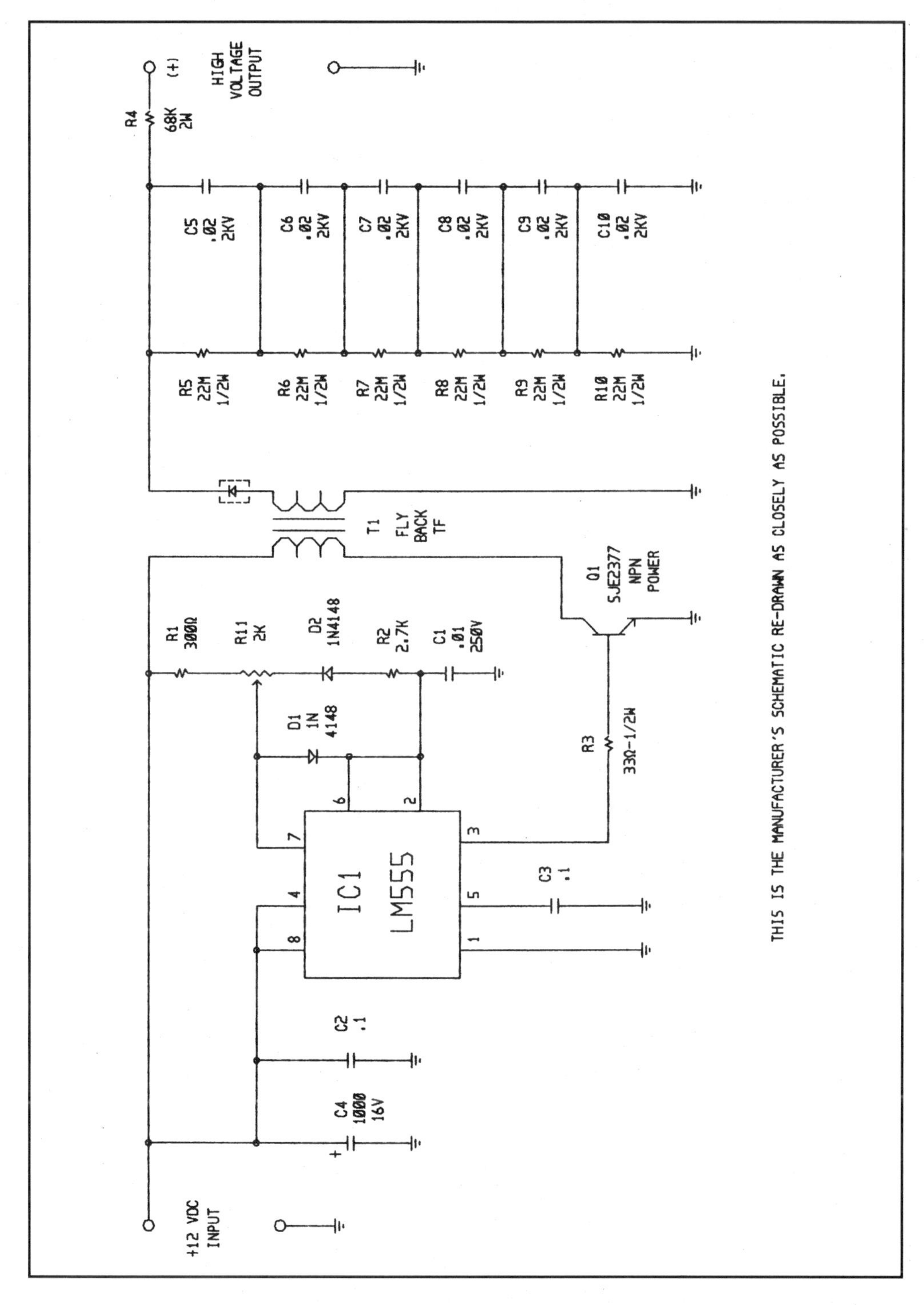

Figure 3-1. Universal helium-neon laser high-voltage power supply.

never know. You don't want to get halfway through the construction only to find that capacitor C8 or diode D3 is missing.

A somewhat weak area with this kit is the instruction sheet. The assembly information is quite brief, and since the PCB is not silkscreened (part locations painted on component side) you want to take care as you place items on the board. There is a parts placement diagram included, but I found it lacking in some details, thus not the easiest to follow.

Also, the location of capacitors C5 and C10 were reversed, as were resistors R5 and R10. Since all components are of the same value, this isn't really a problem. But it does tend to make construction a tad confusing.

Another minor problem is that the PCB location of resistor R2 doesn't coincide with the schematic representation. Here again, the resistor is in series with a diode, and this discrepancy won't affect operation of the supply, but it could be puzzling, especially if you are double-checking the schematic as you install each of the components.

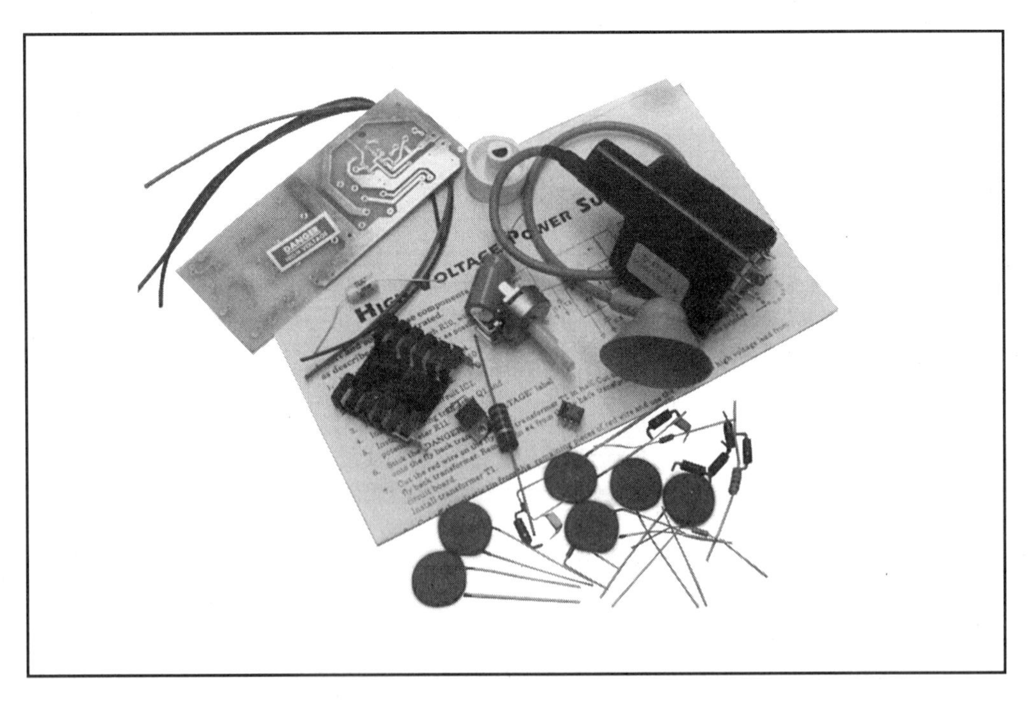

Photo 3-1. The high-voltage power supply kit as it will arrive. All parts are contained in this kit.

The kit doesn't come with an IC socket for the LM555, and I'm a firm believer in IC sockets. So, I added one for IC1. This is not absolutely necessary, but if you have trouble with the circuit, it is nice to be able to quickly replace the 555 to see if that's your problem.

All right. As I do with all kits, I followed the recommended installation sequence per the instructions. That means the resistor and capacitors first, followed by the diodes, IC and transistor. With that done, carefully modify the flyback transformer as outlined in the directions, then solder it to the board. That leaves only the hookup leads. With those installed, the kit is finished, at least in terms of the manufacturer's design. However, I wanted to carry this a little further, concerning both safety and convenience, so let me get to that.

Planning to mount the finished assembly on a plexiglass base and also cover the high-voltage section with a safety plate, I drilled some extra holes in the PCB to accommodate mounting hardware before "stuffing" it. I do recommend these additions, as it makes the supply safer to use and less likely to slide around the workbench.

Referring to *Photo 3-3*, you will see that a small piece of clear 1/16 inch thick sheet plastic is placed over the capacitor/resistor ladder circuitry. This is the high-voltage end of the supply, and the plastic will help prevent your accidentally touching those components. Trust me! You don't want to touch those components when the device is on. Also, don't forget to place the "Danger, High Voltage" label on the plastic shield.

As for the base, it is made from another piece of sheet plexiglass, only this time from 1/4 inch thick stock. The base helps keep you from touching the solder side of the board, protects the solder side of the board and provides some weight for stability.

Additionally, it furnishes a surface (on the right end) to mount an ON-OFF switch and input voltage terminal posts. While neither is an absolute necessity, they do increase the versatility of the completed power supply.

As for the high-voltage output leads, I soldered small car battery-style clips to both the positive (anode) and negative (cathode) wires. These facilitate the metal terminal posts you will find on most laser tubes used with this supply. Be sure the

HV AND 12 VDC POWER SUPPLIES PARTS LIST

HIGH-VOLTAGE SUPPLY

SEMICONDUCTORS

IC1	LM555 Timer/Oscillator
D1, 2	1N4148 Silicon Signal Diodes
Q1	SJE2377 NPN Power Transistor, or Equivalent

RESISTORS

R1	300 Ohm 1/4 Watt Resistor
R2	2,700 Ohm 1/4 Watt Resistor
R3	33 Ohm 1/2 Watt Resistor
R4	68,000 Ohm 2 Watt Resistor
R5-10	22,000,000 Ohm 1/2 Watt Resistors
R11	2,000 Ohm Potentiometer

CAPACITORS

C1	0.01 Microfarad 250 Volt Capacitor
C2, 3	0.1 Microfarad Monolithic Capacitors
C4	1,000 Microfarad Electrolytic Capacitor
C5-10	0.02 Microfarad 2 kilovolt Disk Capacitors

OTHER COMPONENTS

T1	Television Flyback Transformer

12 VDC SUPPLY

SEMICONDUCTORS

U1	7812T 1.5 Amp Positive Regulator
BR1	50 Volt 2 Amp Bridge Rectifier
LED1	Green Jumbo T-1 3/4 Light-Emitting Diode (optional)

RESISTOR

R1	1,000 Ohm 1/4 Watt Resistor (optional)

(Continued next page)

CAPACITORS	
C1	1,000 Microfarad Electrolytic Capacitor
C2	1 Microfarad Electrolytic Capacitor
OTHER COMPONENTS	
T1	120 to 12 VAC/2 Amp Power Transformer
S1	SPST Power Switch
F1	2 Amp Line Fuse
P-	120 VAC Line Cord w/Plug

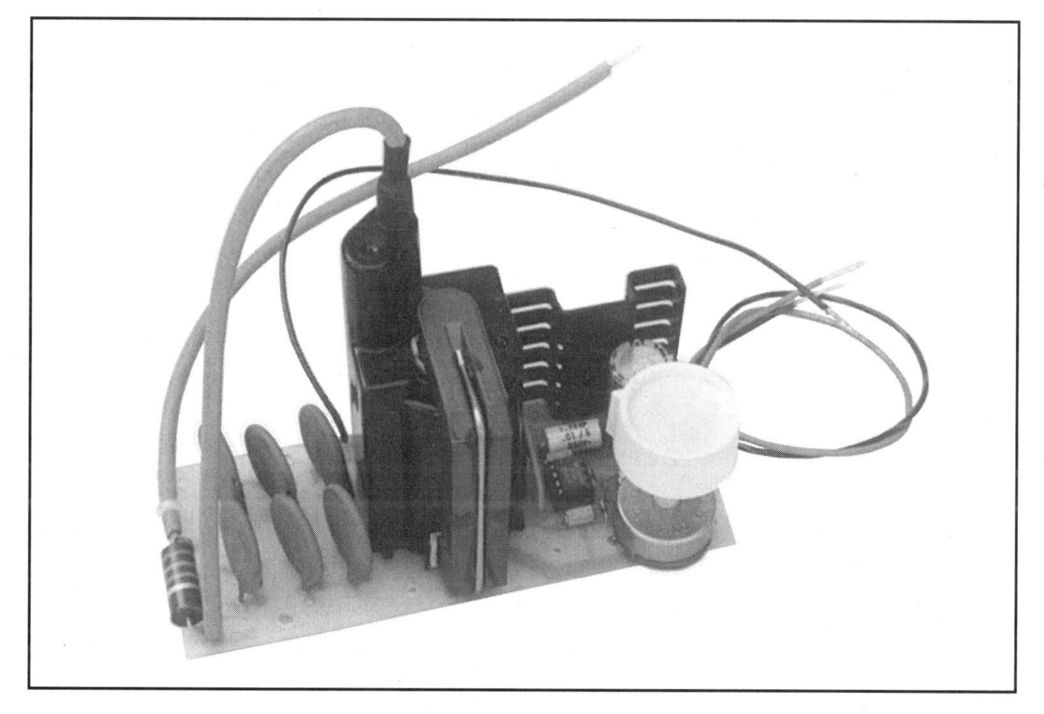

Photo 3-2. The completed high-voltage power supply kit as per the manufacturer's construction directions.

solder connections to the clips are sound; if not, you may experience "flickering" and/or other unstable operation.

That concludes both the kit assembly and the additional safety modifications. The supply is now ready for action, so let's get to the testing and application "stuff."

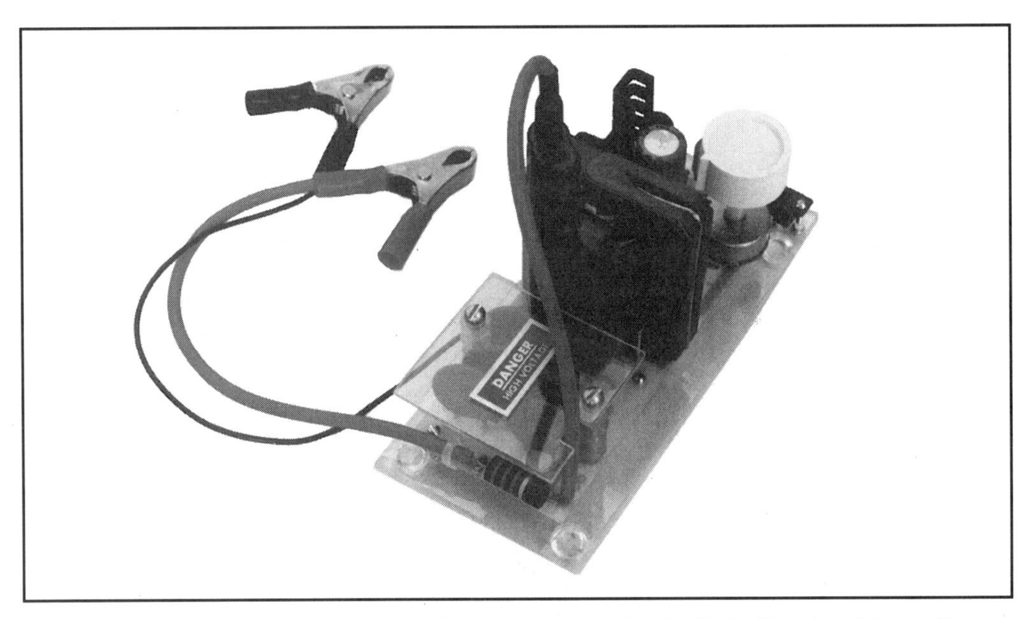

Photo 3-3. The modified high-voltage power suply kit. Note the circuit board is mounted on a plexiglass base and the capacitors are protected by a plastic shield.

TESTING AND OPERATION

The best way to test this supply is to hook up a helium-neon tube rated between 0.5 and 3 milliwatts, set potentiometer R11 to the full counterclockwise position and apply 12 VDC (direct current) to the input voltage terminals. Upon closing the power switch, with most tubes, the bore will glow indicating lasing. However, the tube will probably "sputter" (flicker rapidly) as well, due to insufficient voltage/current.

Next, advance the potentiometer clockwise, and at some point, depending on the laser's rating, the sputtering/flickering will cease. That means the system has reached the proper operating current and will now run smoothly, exhibiting a stable beam.

A second testing method is to actually measure the output voltage and current. This, however, requires access to some high-voltage test equipment that most of us don't have lying around.

If you do, or can find someone to lend you the meters, this supply should provide somewhere between 900 and 2,400 VDC at 3 to 6 milliamps. And, that is roughly the requirements for tubes in the 0.5 to 3 milliwatt range. Naturally, these reading will vary as you adjust potentiometer R11.

Now that you have confirmed proper supply operation, it is time to put it to work. As previously stated, a 12 VDC source is needed, but this can be a large battery pack or any regulated AC-based power supply that furnishes 1 amp of current or more (for 2 to 3 mW tubes, 1.5 to 2 amps may be necessary).

If you would like to construct this supply, I have included a circuit (*Figure 3-2*) that will do the job. It is a standard linear regulated power supply that will yield a "rock stable" 12 VDC and up to 1.5 amps of current. That should be all you will need, except for some of the older 2 to 3 milliwatt tubes.

With your 12-volt source connected to the supply, but not yet turned on, connect the positive high-voltage battery clip to the anode of the laser tube and the negative (ground) clip to the cathode terminal (see *Photo 3-4*). Be sure these connections are sound, but be gentle! HeNe tubes are delicate, and too much pressure on the terminal posts could crack the seals or the envelope itself. Either error will cause the tube to *gas out* (gas mixture escapes) and will ruin the laser.

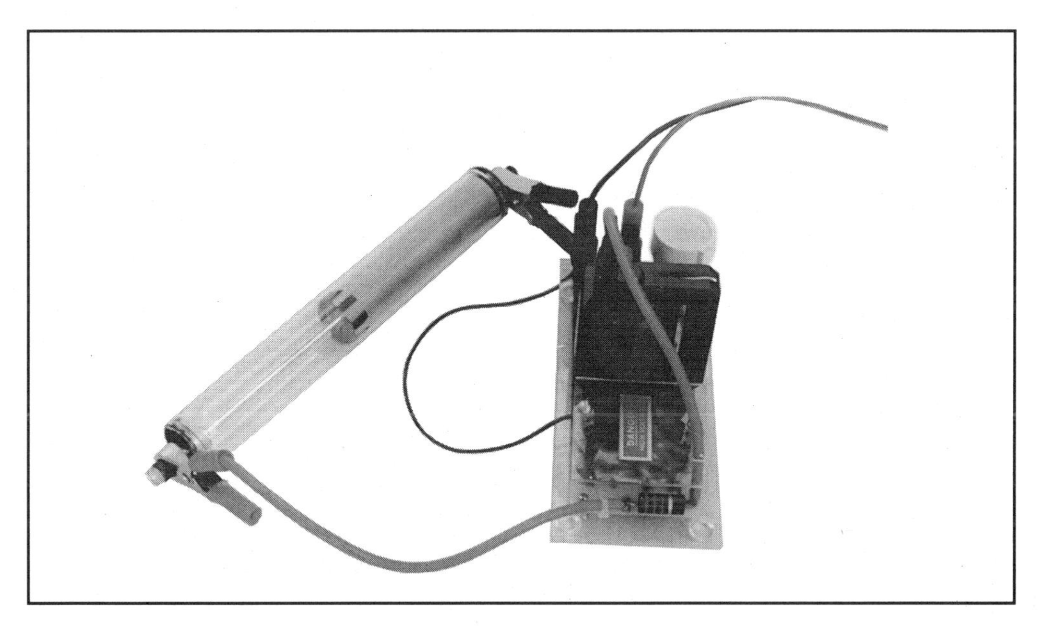

Photo 3-4. The high-voltage power supply kit in operation. Note the glowing bore of the laser tube.

If there seems to be a troublesome connection at the terminal posts, try cleaning them with very fine grade steel wool. These are often made of brass, thus they tarnish easily. However, be VERY CAREFUL with the steel wool as not to scratch the fragile mirrors on the ends of the posts. Any kind of damage to these mirrors can adversely affect the performance of the laser.

OK, if everything is hooked up correctly and ready to go, set the potentiometer to the full counterclockwise position. Next, apply power, and the tube will (should) fire up. Actually, at this setting, the bore will glow, but not consistently. Slowly advance the potentiometer clockwise, and the bore will become much brighter. Eventually you will reach a setting that settles the laser into continuous and stable operation.

The battery-type clips make testing/using different lasers a snap. And, that makes this supply very handy for the workbench. Of course, it can also be the high-voltage source for any laser system you might have in mind. If you decide to enclose the supply in a case, the potentiometer can be panel mounted for easy access.

CONCLUSION

One of the major obstacles with helium-neon laser systems has always been the high-voltage power supply. In the past, they have been hard to find, expensive when you do find them, or just plain unreliable. Epoxy-encased (potted) supplies have become more common and less expensive; thus they are a good investment whenever you encounter them on the surplus market.

But in terms of learning the "nuts and bolts" of how these power sources work, preconstructed supplies are of little help. This supply, however, is especially useful in that respect. By building it yourself, you will gain an appreciation of what makes this part of the laser function and how important it is.

So, have fun constructing the unit and more fun using it. As you trek through the world of lasers, many many good deals (in terms of price) will cross your path regarding helium-neon tubes. With this supply in your arsenal, you will never again have to pass up those deals for lack of an appropriate HV power supply.

Enjoy!!!

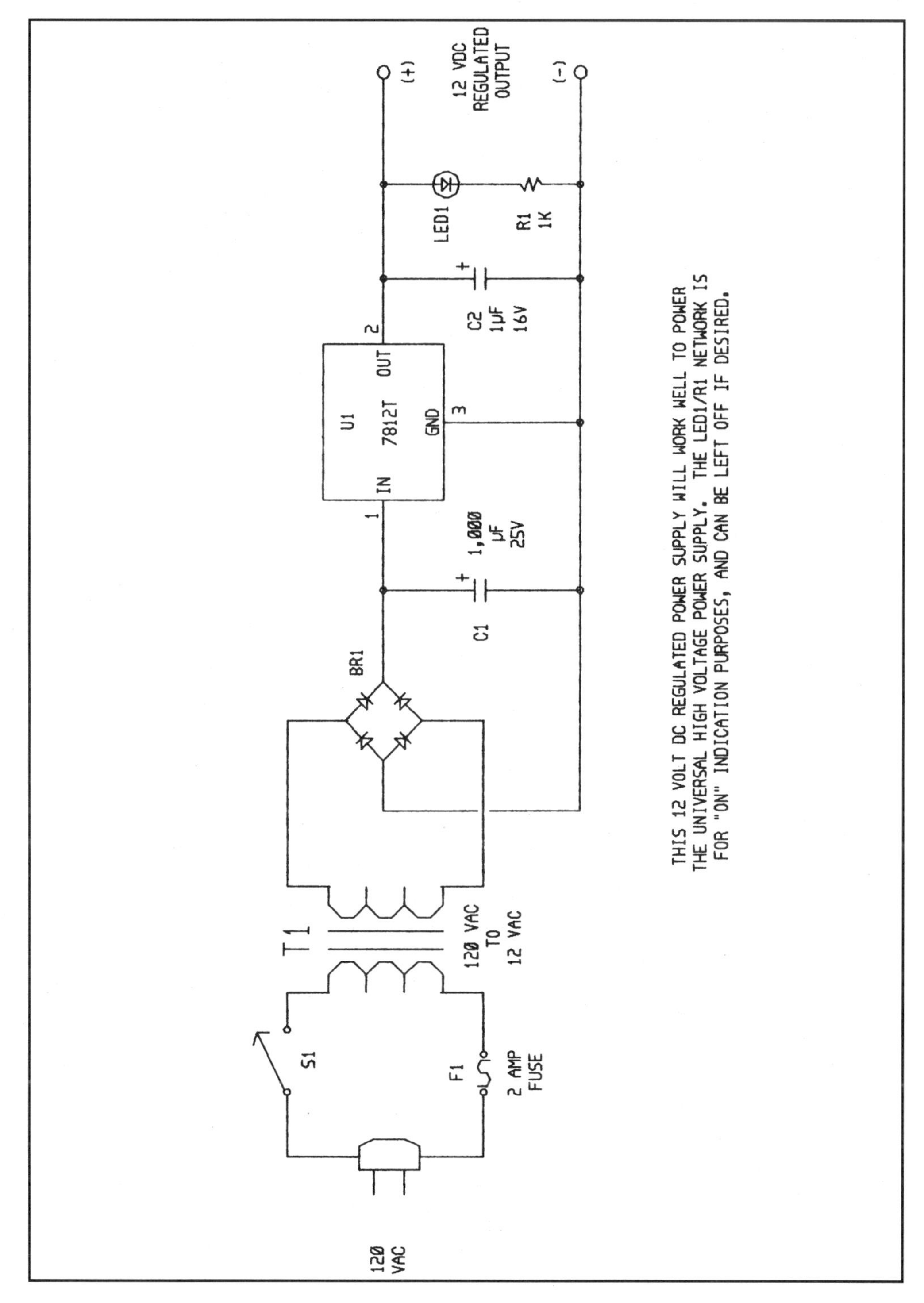

Figure 3-2. 12 VDC regulated power supply.

CHAPTER 4

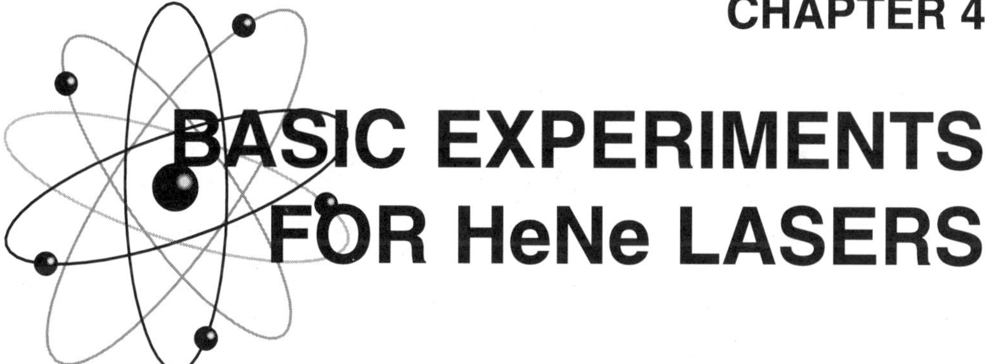

BASIC EXPERIMENTS FOR HeNe LASERS

Now that you have your basic helium-neon system built, it's time to put it to work. In this chapter I will discuss some basic experiments that will familiarize you with the laser's properties. These concern optics, thus providing an introduction to various lenses and other optical devices.

As you probably have already noticed, the helium-neon laser produces a very tight, or *collimated*, beam. However, even that can be improved with the proper lens. Additionally, there are times when you will need to redirect and/or spread the beam or inhibit some of the light, and the world of optics has just the ticket for those tasks. *Photo 4-1* illustrates some of the optics available, but it does not cover everything. Also, you don't necessarily need everything in the photo.

So, without further fanfare, let's examine the fundamental lens types and how they affect light, and also look at some of the other nifty gadgets that will enhance your laser experimenting.

LENSES

Essentially, lenses come in two flavors: *positive* and *negative*. Positive elements condense light, while negative lenses spread it. Both, of course, add their own special touch to the macrocosm of laser light.

Positive lenses are also referred to as *convex*, and negatives bear the handle of *concave*. Furthermore, you can have single convex/concave or double convex/concave, depending on how the lens is designed.

Photo 4-1. A collection of some of the optical elements useful in laser research.
L to R: lenses (single-element and compound); beam splitters;
more single lenses; mirrors and prisms.

Figure 4-1 illustrates the various lens shapes and their respective designations. As can be seen, lenses may be curved inward, outward or both and may also have one flat side. Two flat sides is called a *pane of glass*. Ha!

Convex lenses act to either magnify and/or condense light and concave elements either spread or reduce a beam. The *meniscus* versions also magnify, with the amount of magnification directly dependent on the focal length (the shorter the focal length, the greater the magnification).

In many lens systems, such as those used for photography, astronomy, et cetera, a combination of any or all of these types is employed. Often called *compound* lenses, the individual properties of the various elements work together to achieve a desired result.

OK! Enough about lenses! Let's look at some of the other valuable optical devices.

FILTERS

One way to control the wavelength of light is to filter it; that is, pass the light through a plain piece of glass (or plastic) that is *monochromatic* (a single color). In this fashion, white light will take on the wavelength of the filter.

This, of course, is seen all around us in advertising, traffic lights, rotating beacons, stage productions and the like. If you want red light, use red filter material, or for blue, use blue filters. Combining different filters on separate white light beams will result in "splashes" of color and/or new hues.

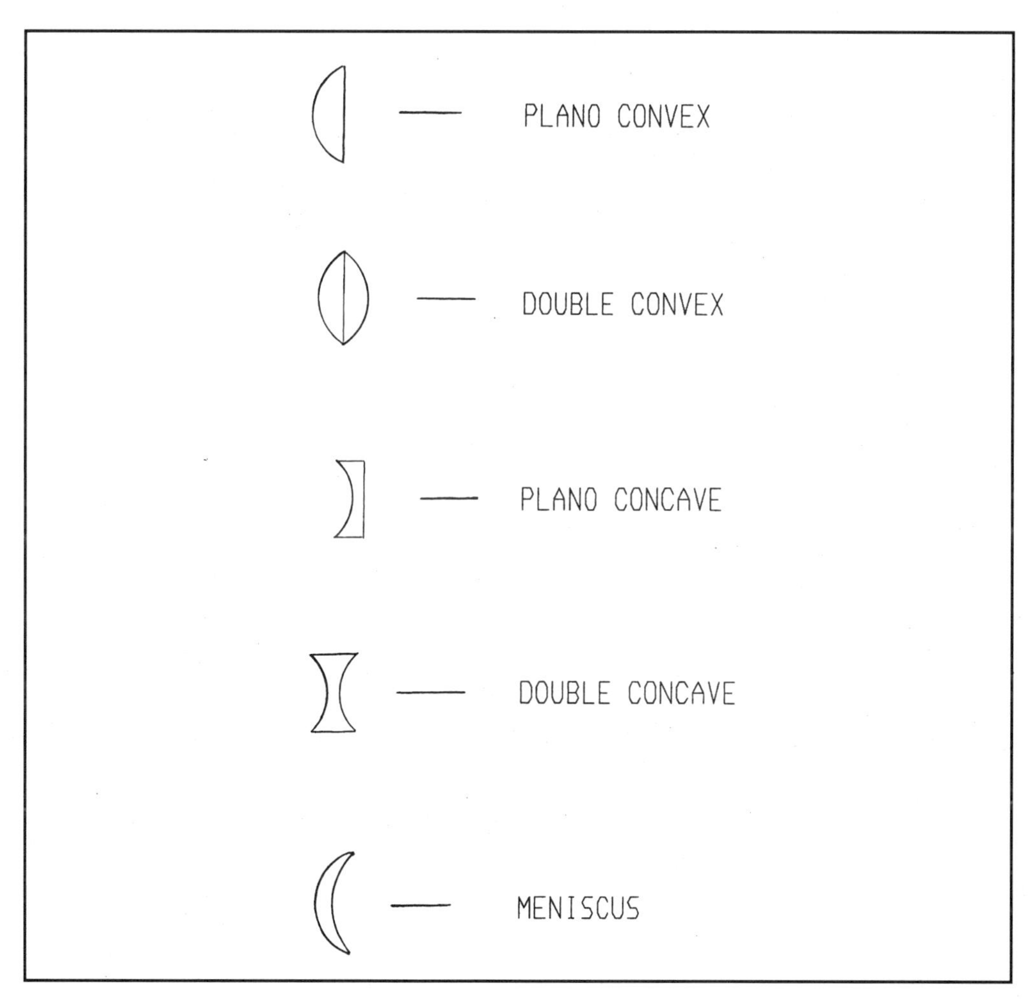

Figure 4-1. Lens shapes/designations.

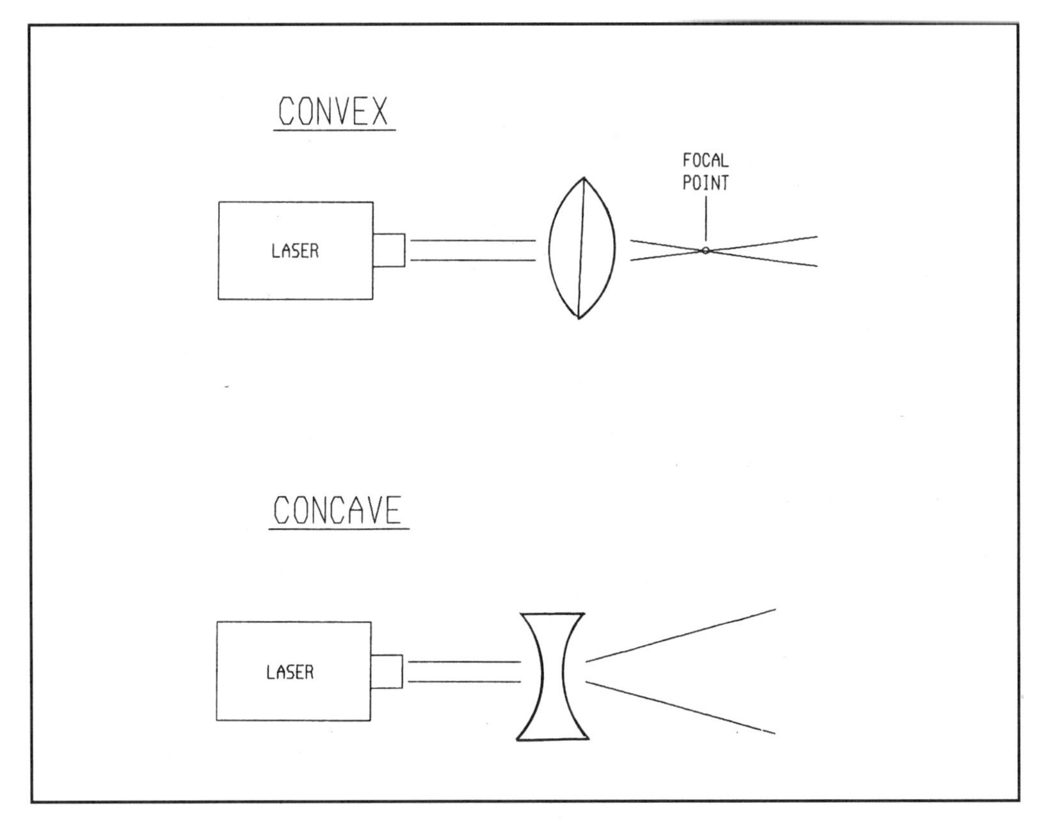

Figure 4-2. Convex/concave lens action.

Another type of filter allows you to reduce reflection. This filter is known as either a *polarizer* or *polar screen*, and is frequently used in photography. However, its properties can be quite effective with laser beams. For example, by polarizing the beam you can easily control its intensity.

A separate filter that also controls magnitude is the *neutral density* filter. These are basically dark filters, often gray in color, that merely prevent some of the light from passing through. These, too, see a lot of service in photography.

One last filter of importance in laser work is the *dichroic*. Here, one wavelength is reflected while all others are passed. Similar to monochromatic filters, these are routinely employed in color television cameras to separate the primary red, green and blue wavelengths. And, the same principle can be applied to multi-laser systems where different wavelengths are encountered.

BEAM SPLITTERS

As its name might imply, a beam splitter "splits" a beam into two separate, but not necessarily equal, parts. You will find 50/50 splitters that reflect half the light while passing the other half. But 30/70, 20/80 and other ratios are also very common. These gems are especially pragmatic in holography and light shows as well as many laboratory applications.

Beam splitters are usually seen in one of two varieties: *partially reflective* mirrors and *prisms*. The first does its job in the same fashion as the partial mirror on one terminal of a HeNe laser, while the second utilizes the reflective/refractive qualities of prisms.

Each has its own special merit, but the *mirror* class is seen most often. This is due to both application requirements and cost. Actually, when it comes to splitting a beam of light, mirrors generally handle the job nearly as well as a prism and are substantially less expensive.

MIRRORS

As for this category, what can I say? Fully reflective mirrors customarily speak for themselves. However, there are a few wrinkles to even this simple a device. For example, mirrors can be either *front-surface* or *back-surface* in concept.

What most of us think of as a mirror is the back-surface variety, where the delicate reflective material is shielded by both the glass and a coat of paint or other protective substance. This type of mirror is quite useful for many laser applications, as well as looking at your pretty face, but you will encounter some distortion due to the beam's back and forth travel through the glass itself.

When such distortion is not acceptable, the front-surface mirror (sometimes referred to as *first-surface*) is your boy. With the reflective surface at the front of the mirror, the light doesn't have to pass twice through the glass "host" while being reflected. Beware, though! These mirrors are very fragile and easily damaged.

Mirrors are also available in either concave or convex shapes. Each provides a special type of reflection, with the concave mirrors condensing the image and the convex expanding it. Most everyone is familiar with those surveillance mirrors in larger places of business that give a wide angle view of the store. This is an example of a convex mirror.

PRISMS

Prisms are an optical field virtually all their own. They come in a multitude of different types, such as *right angle*, *pentaprism*, *equilateral* and so forth. Some are solid pieces of glass or plastic, while others are comprised of two or more elements cemented together.

All have one thing in common: they change the direction of travel of the light that enters them. Many prisms can also divide a beam of white light into the visible light spectrum, from red to violet, forming a rainbow effect.

They are high pragmatic devices to have around a laser lab, but can be quite expensive, especially the glass prisms. However, the surplus market (see source list) is a good place to find both glass and plastic prisms, often with highly reduced price tags.

Again, depending on the type of prism and how it used, they can act as high-performance mirrors and/or beam splitters. You could also make a periscope out of a couple of them if that idea strikes you.

OK, SOME EXPERIMENTS

Now that we've gotten through all that theory stuff, let's have some fun! In an effort to better acquaint you with each of the optical elements just discussed, let's talk about some simple experiments you can perform. Each will, hopefully, demonstrate the capabilities of the device in question. If you don't have a particular optical element, fake it!

As for lenses, I try to keep a large supply of both the negative and positive rascals. Many surplus dealers carry acrylic (plastic) lenses that are available practically for the asking, and for many applications, they do just as good a job as their glass cousins. And, when it comes time to make holographs, a respectable selection will be greatly cherished.

To get started, try shining the laser beam onto a positive lens and notice the fluctuation in the "spot" size as you change the lens's distance from the laser. This variation will not be extreme due to the strong collimation of the HeNe beam, but you should be able to observe some difference. When you want to send that beam over a great distance (such as several miles), positive (convex) lenses will prove outstanding in holding and/or improving the laser's existing collimation.

Next, shine the beam on a negative lens and again vary its distance from the laser. Now you will see a distinct difference in the spot. First, the beam will be *expanded* as opposed to concentrated, and this spreading will vary significantly as the lens position is changed.

As a last experiment, try a combination of both negative and positive elements. Depending on the various configurations you employ, a considerable number of different effects will be observed—all of which will be useful on down the trail.

Now try exposing the red laser beam to some color filters. Red, green and blue are usually the best colors to start with. You will note the beam passes through the red filter with ease, while the blue and green filters inhibit much of the beam. Naturally, the amount of inhibition depends on the density of the filter.

If you have other colors available, try them as well. Normally, as you move away from the red end of the visible spectrum toward the violet wavelengths, the beam absorption increases. However, you might find some surprises along the way.

Beam splitters are next. Position the splitter at a 45 degree angle to the laser and fire it up. (The laser, not the beam splitter!) Depending on the ratio of the splitter, you will behold part of the beam reflecting at a right angle and the remaining light passing on through.

Also, if you have more than one splitter, try splitting the reflected beam a second or third time. Notice how the beam loses strength each time it encounters another splitter. However, in the event you need to send a single beam to a number of individual destinations, and are not restricted by the light loss, this arrangement works right nicely!

Again, beam splitters can be expensive, but the surplus market offers some real bargains, especially regarding the mirror variety. I have paid as much as $7.00 for a single mirror splitter but have also received 3 of them for $1.00. Surprisingly, the cheap ones did just as good a job as that expensive varmint.

When it comes to mirrors, need I say anything? Your best bet is to play around with various types of mirrors just to get a feel for how they reflect laser light. Front-surface mirrors are usually more expensive than the ol' "standby" back-surface gems but, as before, surplus dealers can help you save some coins.

Prisms! Ah yes, prisms! One of the aristocrats of the optical world. As previously mentioned, they come in a variety of different types. And, all can be useful, as they will allow you to direct the beam at will.

Try aiming the laser at one leg (shorter side) of a *right-angle prism* (two legs at 90 degrees) and watch the beam emerge from the hypotenuse (wider side). If the light is exactly perpendicular to the leg, the beam will exit at a 90 degree angle. Changing the entrance angle will vary the exit beam, with a steep perspective usually producing more than one shaft of light.

With *equilateral prisms* (all legs the same width and at an equal angle) you will observe similar activity. However, the exiting beam will usually line up at 60 degrees instead of 90. I say usually because each prism is different; hence they won't necessarily respond the same way.

As for *pentaprisms* (used as viewfinders for single-lens reflex cameras), they receive light from the larger area bottom surface and spit it out the smaller side opening, at a right angle—or vice-versa. Normally the other surfaces are painted flat black and do not allow light to escape.

Incidentally, pentaprisms will reverse the image they receive. This property is not significant with our applications, but with cameras, it does allow the image the lens has reversed to be re-reversed and appear normal in the viewfinder. Just a little trivia for the technically minded.

One last experiment that can be fun is to break the laser beam down into its individual wavelengths. As you may remember, just about all lasers produce several wavelengths, with one being predominant. A right-angle prism is capable of dividing these colors into a spectrum, much as it does with white light.

You will need to do this in a dark area. Shine the beam at the center of the hypotenuse, and the beam will emerge from one leg. However, surrounding the predominant red spot will be other visible light shades being produced by the system. You will have to look closely for these additional hues, as they will be faint, but blues and violets will be most apparent.

CONCLUSION

And, there you have it! A bunch of neat experiments to perform with your HeNe laser and various optical elements. All designed to demonstrate the properties and capabilities of each of the individual categories known collectively as *optics*.

For the "big" experiment—you know, the *grand finale*—try combining each of these specific classifications. You can reflect and split the beam, spread and concentrate it, filter and bend it. Or, in the words of a certain TV commercial, you will have "the whole enchilada." And, each time you change the system elements, a new and unique pattern will evolve.

This will really provide a handle on how to work with and control optics. And, as you proceed down the highway of laser experimenting, this knowledge will be invaluable. Additionally, it's a lot of fun, too! Is that redundant? Oh, well.

So, enjoy this section. It can be a very rewarding learning experience that will also be highly beneficial as you expand your knowledge of lasers. A thorough comprehension of optics, as related to this subject, will add that special edge to each of your laser endeavors—an edge guaranteed to enhance your satisfaction with this fascinating field.

CHAPTER 5

BEAM MODULATION AND 2-WAY COMMUNICATIONS

Like other forms of light, a laser beam can be modulated. In that statement lies the foundation for some very interesting experiments and/or applications. Included here are pattern generators for light shows and "free air" beam communications.

So, these are the areas I will cover in this chapter. One involves "mechanical" modulation, while the other takes an "electronic" approach. Both, however, are pragmatic techniques to achieve laser beam modulation.

As with all modulation, a waveform has to be changed in such a fashion as to convey information. This information can be speech, music, data, images or something else of interest, but in each case, the *vehicle* (waveform) is altered from its conventional form.

This is true of amplitude or frequency modulation (AM/FM) of radio waves for sound, carrier waves (CW) for code and radio teletype (RTTY) for data. And, the same applies to either standard white light (noncollimated) or laser light, where the vehicle is the beam.

So, without further fanfair, let's take a look at a couple of methods that will do the trick with your laser. Hey, you'll like 'em! Trust me!

MECHANICAL APPROACH THEORY AND CONSTRUCTION

This one could easily have ended up in the next chapter on light shows, but instead, I decided to include it here, mainly because this device actually modulates the beam. It may do so in a relatively mundane fashion, but it modulates nonetheless.

Well anyway, what I'm about to describe is a lot of fun to play with. It will produce some fascinating designs that are directly correlated to whatever signal is introduced. And, it is extremely simple to build. I mean, you just wait and see. This gem does wonders with just two components.

OK, let me characterize this project in more detail. What you need is an audio style speaker and a small mirror. Now, pay close attention, as this is the hard part. You take the mirror in your right hand, apply some cement to its back, and glue the mirror to the surface of the speaker's cone.

I know this is tough, but with some patience, and maybe some practice, you will master it in no time. Naturally, I'm being facetious. Oh, don't forget to let the cement dry before you test the device—otherwise, you might just have to start all over.

I know, I know! Enough! This is no longer funny! So, I will get back to business. The purpose of gluing a mirror to the cone of a speaker is to make said mirror vibrate when said cone does...or, something like that.

In this manner, when music, voice or other sounds are sent to the speaker from an amplifier, radio, et cetera, the mirror will quiver in unison with those sounds. And, when you aim a laser beam at the mirror, the reflected beam will dance enthusiastically, off a nearby wall or ceiling, displaying circles and leaping lines and other dramatic patterns.

The completed speaker/mirror assembly should look something like *Photo 5-1* and can be mounted to a stand if desired. Usually, the larger the speaker, the better the reflected pattern, although small speakers do a darn respectable job.

Another trick that will help increase the overall movement of the mirror is to glue it first to a small, relatively flexible spring, then to the speaker cone. If you will excuse the pun, this puts a spring in the mirror's step.

MECHANICAL APPROACH CONCLUSION

With this piece of equipment in hand, you will be able to entertain the kids, the cat or an awfully "slow" adult for hours on end. It works especially well with music,

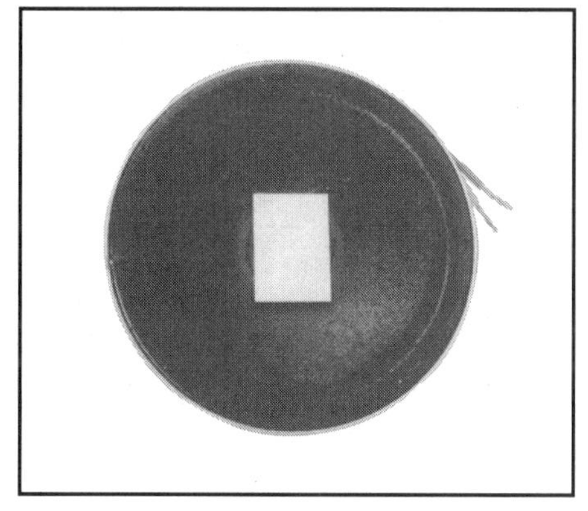

Photo 5-1. Closeup shot of the very simple "mirror glued to a speaker" laser light modulator.

as the vigorous patterns keep in step with the harmony. As I said earlier, this device yields a simple but dang blame good light show.

It won't induce much of a burden in terms of toil or cost—I mean, considering all the fun you will have—so give it a try. This unit will come in handy more often than you might think.

ELECTRONIC APPROACH THEORY

This system has a little more meat to it. Here, we will literally change the current level being fed to the helium-neon laser tube. The variation is subtle, so subtle that your eye will probably not be able to detect it. But, to the electronic sensor receiving the signal, it will be more than sufficient.

Sufficient? Sufficient for what? Well, sufficient to transmit a variety of different information over the laser beam. To explain this a little further, the sound introduced to the system will cause the intensity of the laser to vary according to the intensity of said sound.

However, since lasers produce such a powerful light to start with, minor changes in the magnitude are difficult to see. When you are dealing with regular light sources (a flashlight for example), these variations are quite apparent. And, that makes the principle behind this type of modulation easier to comprehend.

But, the light wave receiver described in this section is more than capable of discerning those minute fluctuations. And, with the aid of external optics, the receiver can detect the beam at surprisingly long distances, with astonishing clarity.

OK, with that said, let's look at the transmitter, or laser modulator, first. Examining *Figure 5-1*, you will see that the amplifier output is connected to the secondary of a *filament* transformer. The transformer primary is then wired in series with the cathode (negative) line separating the high-voltage power supply and the laser.

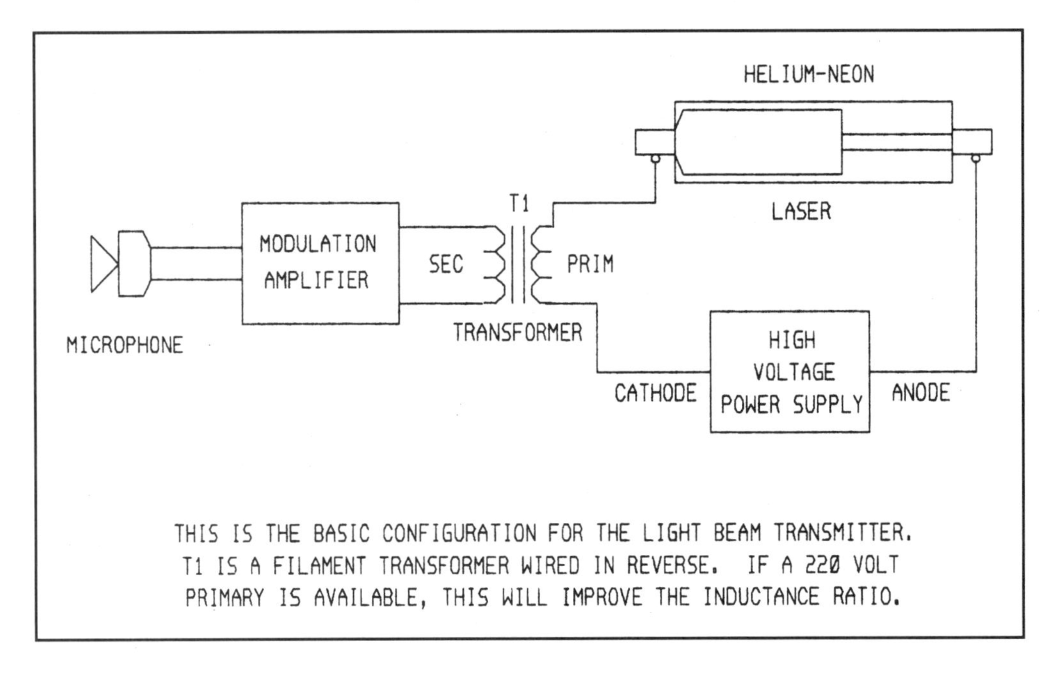

Figure 5-1. Light beam transmitter block diagram.

With this arrangement, the amplifier output controls the current passing through the laser by means of induction between the secondary and primary windings. As the amplifier output changes, so does the magnetic field in the secondary. This, in turn, proportionally modifies the amount of current sent to the laser by the transformer's primary.

The reverse wiring of the transformer (secondary to primary), is employed because this *turns ratio* produces a greater variation in the beam's intensity; hence, a signal that is easier to decode by the receiver.

While virtually any audio amplifier will work, I have included the schematic for a simple, yet effective integrated circuit (IC) configuration that does a dandy job (see *Figure 5-2*). Here, an LM386 amplifier chip is fed with a preamplifier and terminated with the transformer, allowing whatever the microphone hears to be transmitted over the beam.

The preamp consists of transistor Q1, resistors R2 and R3, and capacitors C1 and C2. C1 couples the audio from the microphone (MIC), which is powered by resistor R1 and the positive voltage, while resistors R2 and R3 set bias for the circuit.

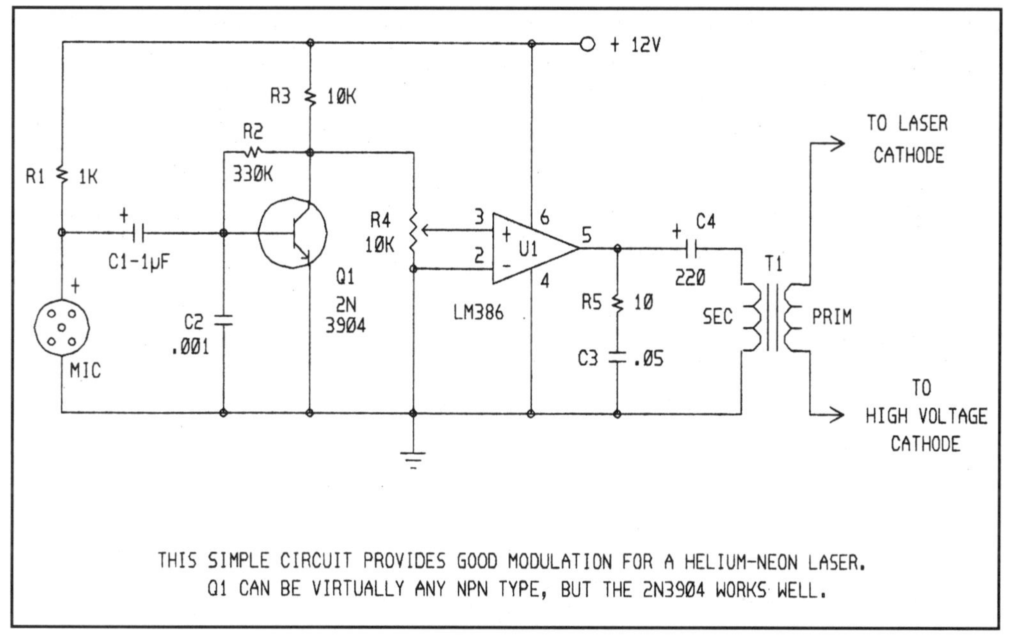

Figure 5-2. Light beam transmitter schematic.

Capacitor C2 acts to shape the audio frequency response of the preamp, as does C1 to some extent. Here, the values chosen for the capacitors tend to mute the high-frequency tones, giving the amp a more "basic" sound.

As for the main, or power amplifier, potentiometer R4 functions as a volume control by limiting input to the LM386 (U1). As can be seen, U1 internally handles most of the work, but resistor R5 and capacitor C3 are included to balance the IC's output regarding the load—in this case the secondary of a power transformer. C4 is added to couple the output to the load.

As previously mentioned, transformer T1 may be any *filament* unit as long as it can manage the load of the laser system. Actually, in the past, I have used audio transformers for this purpose, and they have worked just fine. However, a little heavier duty inductor is probably a sound idea. Thus, here I have recommended a filament transformer.

Also, if the primary is rated for 220 to 240 volts, so much the better. The higher ratio of turns between primary and secondary will definitely improve the transfer of signal from the amp to the laser.

OK, that takes care of the transmitter—now we have to consider a receiver. *Figure 5-3* is a block diagram for one receiver system, but others will surely do the job. This is basically nothing more than an amplifier with a phototransistor at the input.

When the modulated light strikes the phototransistor, it varies the input to the amplifier in accordance with the magnitude of the beam. As the beam fluctuates, so does the amp's output, and this constitutes the signal decoding. Finally, all of this resolves in an announcement of the signal over the speaker.

Figure 5-4 represents the schematic diagram for the receiver I pictured in *Figure 5-3*. Notice the similarity between this circuit and the transmitter's modulator amp.

In fact, the only real difference is the phototransistor on the input and the speaker on the output. The rest of the circuit is the same preamp/power amp combination and operates in concurrence with the above description.

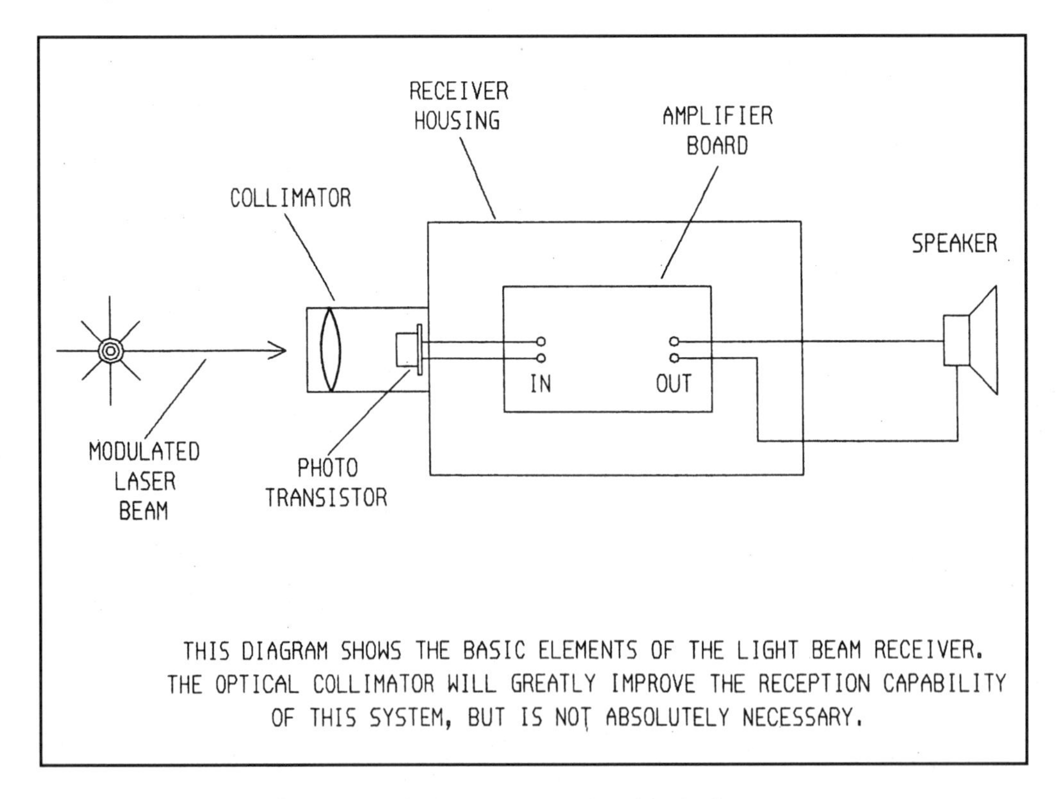

Figure 5-3. Light beam receiver block diagram.

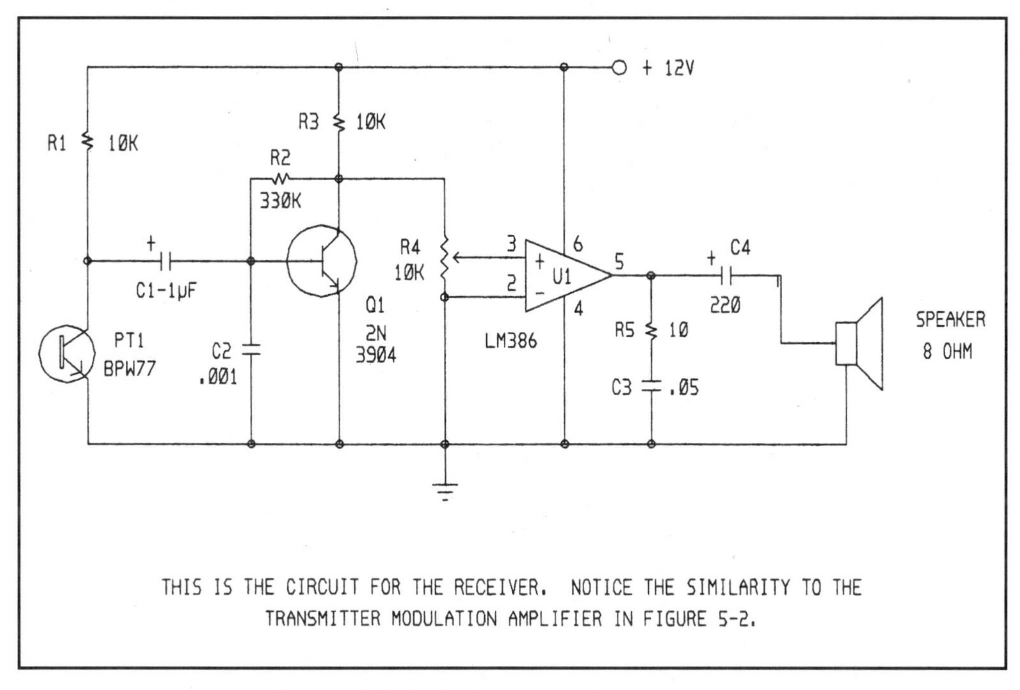

Figure 5-4. Light beam receiver schematic.

Another way to work this is to use a small "preassembled" amplifier. These are sold by a number of electronics outfits, including Radio Shack, and have the amp board and speaker built into a small case. By connecting a *solar cell* (solar battery) to the input, instead of a phototransistor, the whole unit makes a dandy audio receiver.

So, that's the theory. Now, let's look at how to put this gem together and use it.

CONSTRUCTION AND OPERATION

Photo 5-2 illustrates a "breadboarded" transmitter that I used as the prototype, while *Photo 5-3* shows the receiver. This approach did well for my purposes, but perf-boards or printed-circuit boards (PCB) can easily be employed if a permanent arrangement is desired.

The layout of these circuits is hardly critical, but as with any audio device, "clean" wiring helps keep the systems quiet. Try to avoid long hookup lines that tend to pick up parasitic signals, and use a solid (substantial) ground rail.

LIGHT BEAM COMMUNICATIONS PARTS LIST

LIGHT BEAM TRANSMITTER

SEMICONDUCTORS

U1	LM386 Low-Power Audio Amplifier
Q1	2N3904 NPN Signal Transistor
MIC	Electret Style Microphone

RESISTORS

R1	1,000 Ohm 1/4 Watt Resistor
R2	330,000 Ohm 1/4 Watt Resistor
R3	10,000 Ohm 1/4 Watt Resistor
R4	10,000 Ohm Panel Mount Potentiometer
R5	10 Ohm 1/4 Watt Resistor

CAPACITORS

C1	1 Microfarad Electrolytic Capacitor
C2	0.001 Microfarad Mylar Capacitor
C3	0.05 Microfarad Mylar Capacitor
C4	220 Microfarad Electrolytic Capacitor

OTHER COMPONENTS

MIC	Electret Condenser Microphone
T1	Filament Style Transformer (120/220 VAC to 12VAC)

LIGHT BEAM RECEIVER

SEMICONDUCTORS

U1	LM386 Low-Power Audio Amplifier
PT1	BPW77 NPN Phototransistor

(Continued next page)

RESISTORS

R1	10,000 Ohm 1/4 Watt Resistor
R2	330,000 Ohm 1/4 watt Resistor
R3	10,000 Ohm 1/4 Watt Resistor
R4	10,000 Ohm Panel Mount Potentiometer
R5	10 Ohm 1/4 Watt resistor

CAPACITORS

C1	1 Microfarad Electrolytic Capacitor
C2	0.001 Microfarad Mylar Capacitor
C3	0.05 Microfarad Mylar Capacitor
C4	220 Microfarad Electrolytic Capacitor

OTHER COMPONENTS

SPKR	8 Ohm Speaker

MISCELLANEOUS

Solder, PCB Materials, Hardware, Cases (if desired), Collimator Materials (if used), IC Sockets, Hookup Wire, 12 VDC Power Supply, Etc.

When a PCB is employed, try to retain on the board as much copper foil as possible and tie it to ground. This will greatly aid the system in rejecting that annoying, irritating, inflammatory noise that's all around us.

Another trick, if you are using a case, is to shield the enclosure by gluing metal foil to the inside surfaces. This shield should then be connected, in some fashion, to ground. Here again, extraneous signals will be intercepted by the shield and summarily sent to ground where they can do you no further harm. And that's where they belong! So there!

On the receiver, it is wise to incorporate an optical system (collimator) for the phototransistor. This will concentrate the incoming beam on the transistor, mak-

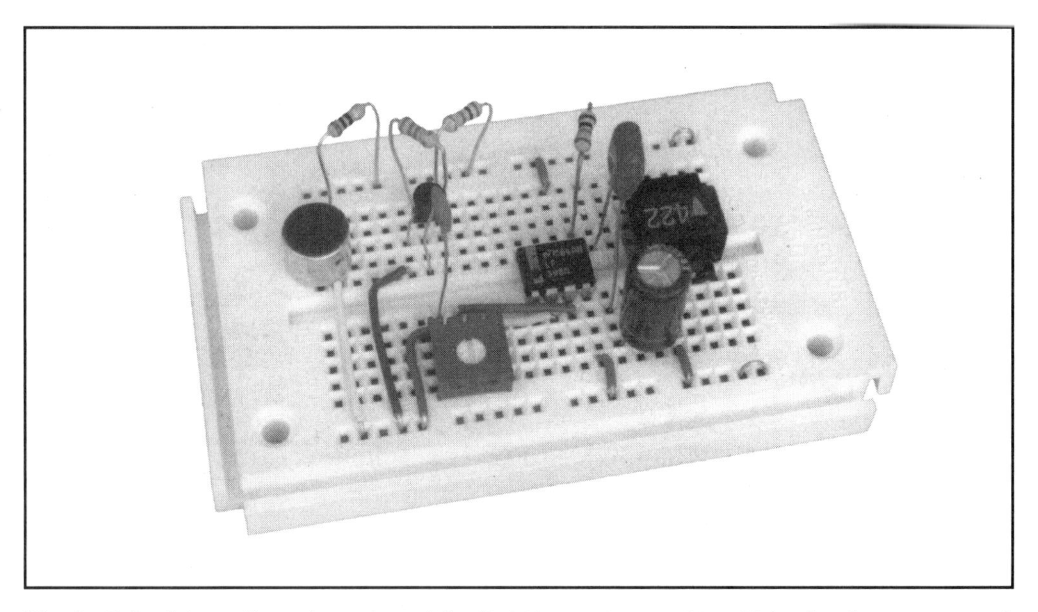

Photo 5-2. A breadboard version of the light beam transmitter. This circuit uses a small "audio" transformer, but you may well find a larger filament type does a better job.

ing reception better. It can be made from a magnifying glass and isn't an absolute necessity. But, it sure does help.

Anyway, that pretty much covers the construction of these systems. This isn't rocket science, so they aren't difficult. Using the sound construction techniques that I'm sure you always use should prove most satisfactory.

On that note, why don't we talk about using this marvel of modern electronics? Well, you will be happy to learn that this ain't rocket science either. In fact, it is quite simple.

Since we are modulating a visible light beam (I'm presuming you are modulating a visible beam), it is easy to see where the light falls. With some of the more clandestine systems, infrared lasers are employed, and this leads to the necessity for an aiming mechanism.

Alright, back to business. The transmitted beam merely has to strike the receiver's collimator, or phototransistor if you have not included a collimator. In this fashion, whatever "data" you are sending over the beam will be decoded by the receiver at the other end. And, viola!...light wave communications!

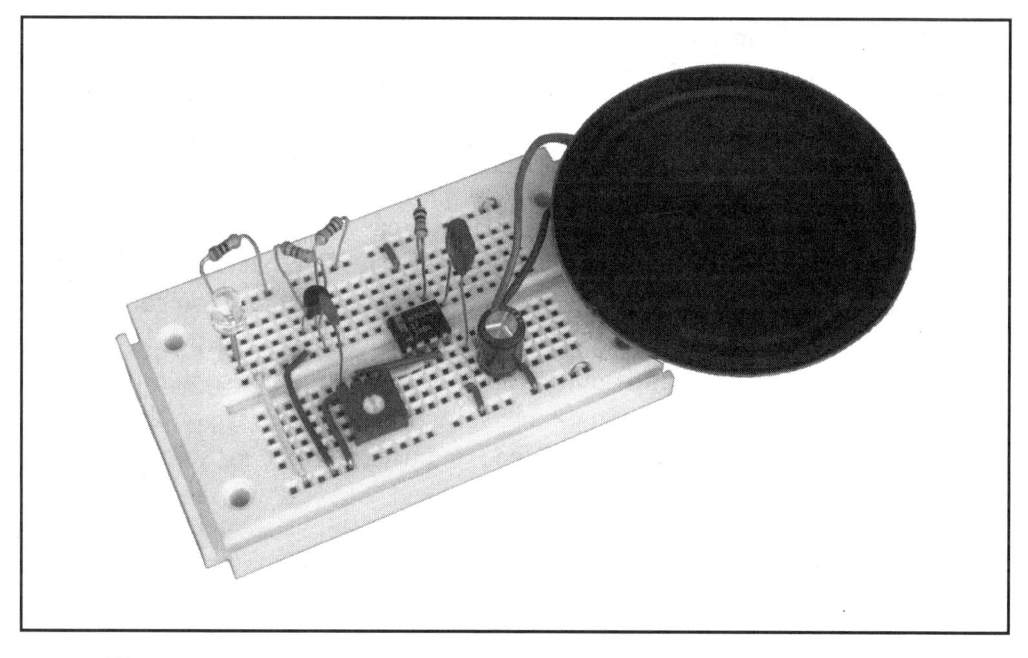

Photo 5-3. A breadboarded version of the light beam receiver. As seen, this is a relatively simple circuit.

Naturally, if you wish to carry on a "2-way" conversation, you will need two transmitters and two receivers. Place a pair of them at each end of the free air link, and you have your duplex system. Kind of like a telephone, only better!

One big advantage to light communication is that unlike radio signals, it is virtually immune to external interference. Ambient light can cause some trouble in bright surroundings, but that is easily corrected by shielding the sensor.

However, a downside to light is that you have to be able to see it. If something should happen to get in between you and the receiver, you link is instantly broken. Usually, though, a little careful planning regarding transmitter/receiver location will avoid such a disaster.

CONCLUSION

Of the many applications of a laser, this one has always been fascinating. The prospect of being able to communicate with light is almost as amazing as being able to communicate with RF. At least in my mind, anyway.

Also, the clarity of those communications is really remarkable. Having been involved in some form of radio frequency communications most of my "electronically aware" life, I find it difficult to describe my astonishment at the quality of light-based signals—especially when a laser is employed.

The equipment is relatively simple and CHEAP, easy to use, and the results are highly rewarding. I mean, how can you resist this stuff? What a project! What a way to spend a weekend! What a way to....well, enough of that. You get the idea!

Just try this one, and you'll find out for yourself. It's guaranteed to impress everyone. Well, kind of. So, go for it and have some fun. Besides, if you don't like it you can always reconnect the cathode lead and use the laser for something else. See, I try to please everybody!

CHAPTER 6

A LASER LIGHT SHOW

This project is just plain fun! Fun to build and a whole lot of fun to play with. Other attributes include being easy and CHEAP to construct. I know, those are relative terms. But, *relative* to the electronics hobby, they do apply.

What we have here is a system that allows you to control the speed of three separate motors. When you attach round mirrors to the shaft of each motor and align them properly, a laser beam will reflect off one mirror, then the next, and finally the third to emerge as a pattern.

The patterns are controlled by the speed of the individual motors and are virtually unlimited in terms of shape and novelty. As you change the controls, even slightly, the lissajous design generated by this machine will also change. And, the results can be spectacular!

To further enhance the show, the mirrors are mounted to the motors at a slight angle and slightly off center. This provides a "quiver" effect that embellishes the normal beam reflection.

So, gather up your materials, as itemized in the parts list, and give this one a try. It will provide you and others with countless hours of fascinating entertainment. I'll tell you, the "kids" will love it!

THEORY

One of the best ways to control the speed of a direct current (DC) motor is to power it with a pulse-width modulated source. In this fashion, the *duty cycle* of the signal regulates the ON/OFF time of the motor, and thus, its speed.

Using this approach, as opposed to a potentiometer, will keep the speed stable because the pulse-width generator will maintain a constant output even when the supply voltage changes. With the potentiometer scenario, the voltage to the motor will vary in accordance with changes in supply voltage; hence, the speed will also vary.

So, how do we do this? As with many electronic tasks, there is more than one way to accomplish the job, and this project is no exception. But each solution does have one aspect in common, and that is to control motor speed through pulse-width modulation.

Pulse-width modulators can be constructed from transistors, gates or dedicated oscillator integrated circuits. And, each approach results in a square wave oscillator output. Naturally, a square wave signal gives the most distinct changes in the ON and OFF states, and provides the best motor regulation.

With these options available, I chose to go the dedicated oscillator route, using an old friend, the LM555. Referring to *Figure 6-1*, you will see a relatively (there's that word again) simple circuit comprised of a voltage regulator and three identical 555 oscillators.

The 7812T regulator (U1) is configured as a standard linear arrangement where C1 handles filtering of the input direct current (DC), and C2 removes transient signals and prevents chip oscillation. The R1/LED1 network is merely an ON indicator.

The purpose of this addition to the circuit is to provide a stable voltage from which to operate the timers. In this fashion, the 555 output will stay constant, hence the motor rotation speed will not change even if the input voltage changes.

Integrated circuits (ICs) U2, U3 and U4 (and friends) comprise the three 555 oscillator sections, with each wired in the "astable" mode. The frequency of these circuits is controlled by the potentiometer (R2, R4 or R6), the fixed resistor (R3, R5 or R7) and the capacitor (C3, C4 or C5). The values chosen in this case provide a good range of motor speeds.

Speaking of the motors, they are connected between the 555 output (pin 3) and ground. I employed small hobby-style DC motors, and they functioned quite well

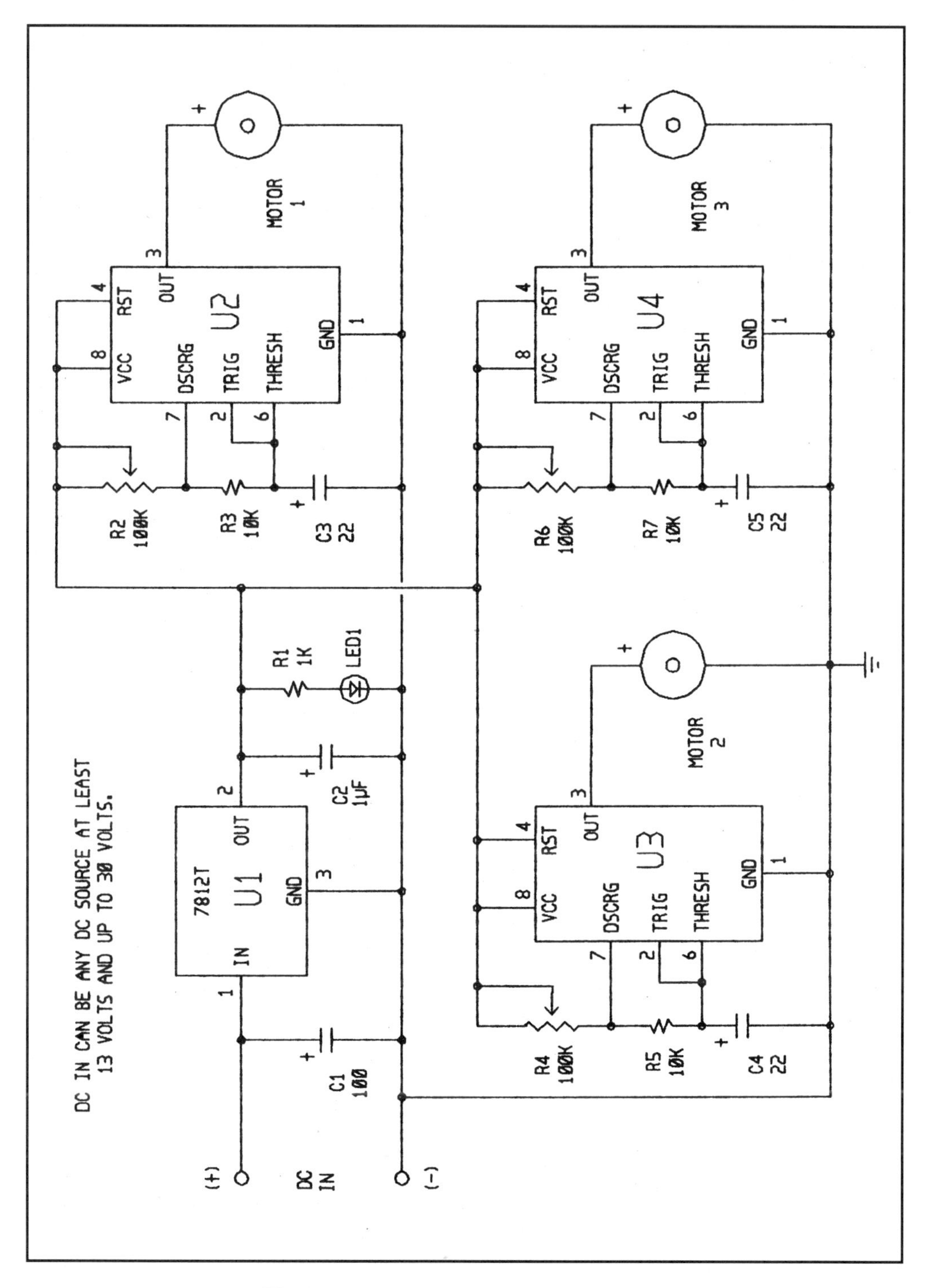

Figure 6-1. Laser light show schematic.

directly off the oscillators. However, different motors require different current levels (How's that for stating the obvious?), and you may need to employ the simple transistor driver circuit in *Figure 6-2*.

If the motors you select (which by the way, should all be the same type) want to run sluggishly (or not at all), most likely the problem is too little current. The single transistor drivers will quickly remedy this problem. Note: I designated the transistor as the TIP31, but virtually any NPN power transistor will do the job.

And that, as they say, is that. Nothing particularly complicated, but this arrangement does produce a very stable pulse-width modulated DC motor control system.

Now, let's look at the mechanical theory, such as it is. As mentioned in the Introduction, each of the three motors has a small round mirror mounted on its shaft. They provide a *reflection path*, and rotation of these mirrors, which are positioned slightly out of kilter, results in a circular/semicircular pattern of reflection.

When you add adjustable rotation speed to the equation, the beam reflection becomes even more distorted, and the ensuing design reveals that alteration. The variation that can be achieved, of course, is practically limitless. It will largely depend on the setting applied to each of the three motors.

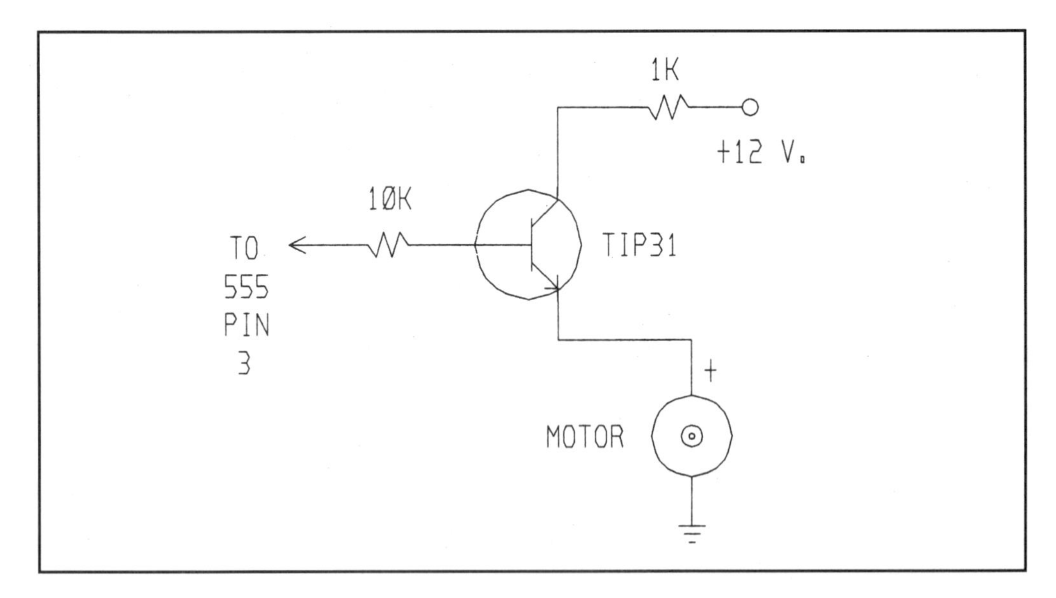

Figure 6-2. Motor driver circuit.

CONSTRUCTION

You can start with either section of this project, but I went with the electronics first. I am a BIG fan of printed-circuit boards (PCBs), and in keeping with that preference, I built this circuit on a PCB. If you would like to do the same, I have included the foil pattern in *Figure 6-3*. Also, *Figure 6-4* provides parts placement. However, perf-board and point-to-point wiring will work if you feel more comfortable with it.

With the PCB in hand, first install the IC sockets, then the small discrete components (resistors, capacitors, etc.). I'm also a firm believer in IC sockets! Next comes the regulator (U1) and the LED. As always, observe polarity and orientation on any component vulnerable to such things.

Now, solder the three potentiometers to the appropriate board pads. These can be mounted to a "standalone" panel, as in the prototype, or somewhere on a case if you elect to use an enclosure. Either way, be sure to consider the potentiometer locations when determining the length of hookup wire.

Our next step is to construct the motor assembly. In the prototype, I mounted the three motor/mirror units to a plexiglass base, which also incorporates the speed control circuitry. That keeps everything together, but depending on your anticipated needs, the two sections can be separate.

First, let's talk about installing the mirrors on the motors. This procedure does require a little care, as you want to mount the mirrors slightly off center and at a slight angle to the motor shafts. However, you don't want to overdo the program or you might find part of the beam spilling off the edge of the last mirror.

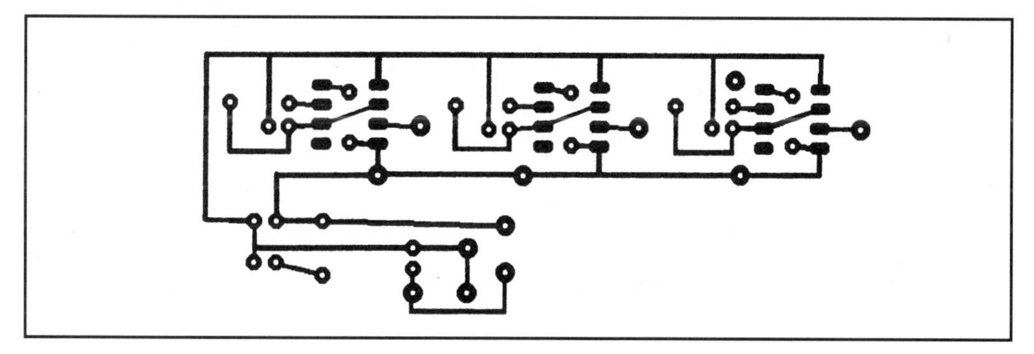

Figure 6-3. Laser light show PCB foil pattern.

LASER LIGHT SHOW PARTS LIST

SEMICONDUCTORS

U1	7812T Positive 12 Volt Regulator
U2-4	LM555 Timer/Oscillators
LED1	Jumbo T-1 3/4 Light Emitting Diode

RESISTORS

R1	1,000 Ohm 1/4 Watt Resistor
R2, 4, 6	100,000 Ohm Panel Potentiometers
R3, 5, 7	10,000 Ohm 1/4 Watt Resistor

CAPACITORS

C1	100 Microfarad Electrolytic Capacitor
C2	1 Microfarad Electrolytic Capacitor
C3, 4, 5	22 Microfarad Electrolytic Capacitors

OTHER COMPONENTS

M1-3	6 to 12 Volt Hobby Motors

MISCELLANEOUS

PCB Materials, Solder, Hook-Up Wire, Plexiglass, Cable ties, Hardware, DC Power Supply, Etc.

I found that a 3-degree angle about 1/8 inch off center did nicely. This arrangement produces a "goodly" amount of "wobble," if you will, but regarding your application, some experimenting might reveal a more satisfactory result. Also, you don't necessarily have to have all mirrors mounted exactly the same. A little variation might well help the effect.

To attach the mirrors, I contact cemented 1/2 inch thick plexiglass disks with holes drilled slightly smaller than the diameter of the motor shaft. This provided both a secure link for the mirrors and a snug fit for the shafts.

But, wood could also be used in place of the plexiglass. I would advise securing the motor shafts to the wood with epoxy. This will prevent the mirrors from loosening up after prolonged use and possibly flying off. That could be disastrous, as well as dangerous, with glass mirrors.

Speaking of the mirrors, I haven't yet elaborated on this component of the system. For some good news, these mirrors can be either front- or rear-surface, any size between about half an inch and 3 or 4 inches, and can be made from glass or plastic.

For some bad news (well, not really so bad), round mirrors can be a bit of a challenge to find, especially in the size(s) we need. I located a double handful of 3-inch round glass jobs at American Science and Surplus, as seen in *Photo 6-1*, and they worked very well. But, being surplus, they may not still be available at the time you read this. Check the source list for other possibilities.

Also, here are a few other places to look. Seen quite frequently on the surplus market are dental inspection mirrors. These are usually in the 3/4-inch round size and will fit the bill, provided you don't attach them too far off center.

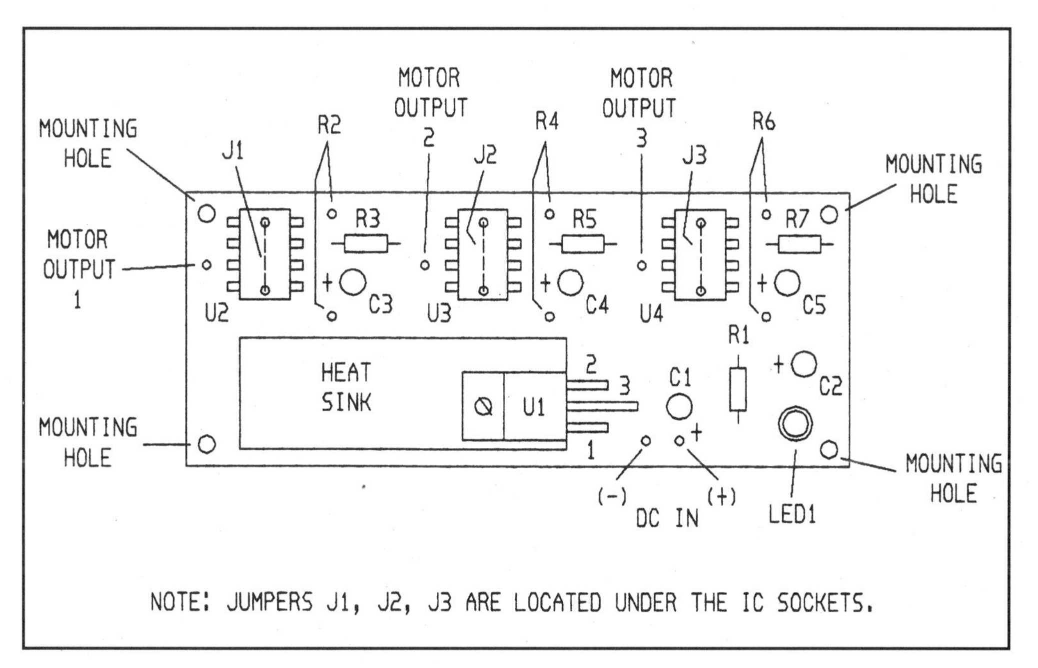

Figure 6-4. Laser light show parts placement.

You local drug, discount or dollar store is a great place to look for small round "cosmetic" mirrors. These are made primarily for carrying in purses, or pockets, to accommodate a quick appearance check (that's personal appearance, of course), and again will do a good job for this purpose.

Most plastic supply houses carry plastic mirror material which in addition to reflecting the beam very well, afford a degree of safety; that is, they do not break as easily as their glass cousins. However, I have run into some difficulty in finding companies that have the capability to cut the plastic into small disks. So, unless you can do this yourself, the "plastic houses" may not be a pragmatic source.

And, that gives you several possibilities regarding the mirrors. Look around, and I imagine you will encounter others. On a last "mirror" note, the mirrors don't absolutely have to be round. If all else fails, square units can be employed, but the imbalance created by rotating squares could place undue wear and stress on the motors. Hence, I wouldn't use anything but round mirrors unless you're left with no other choice.

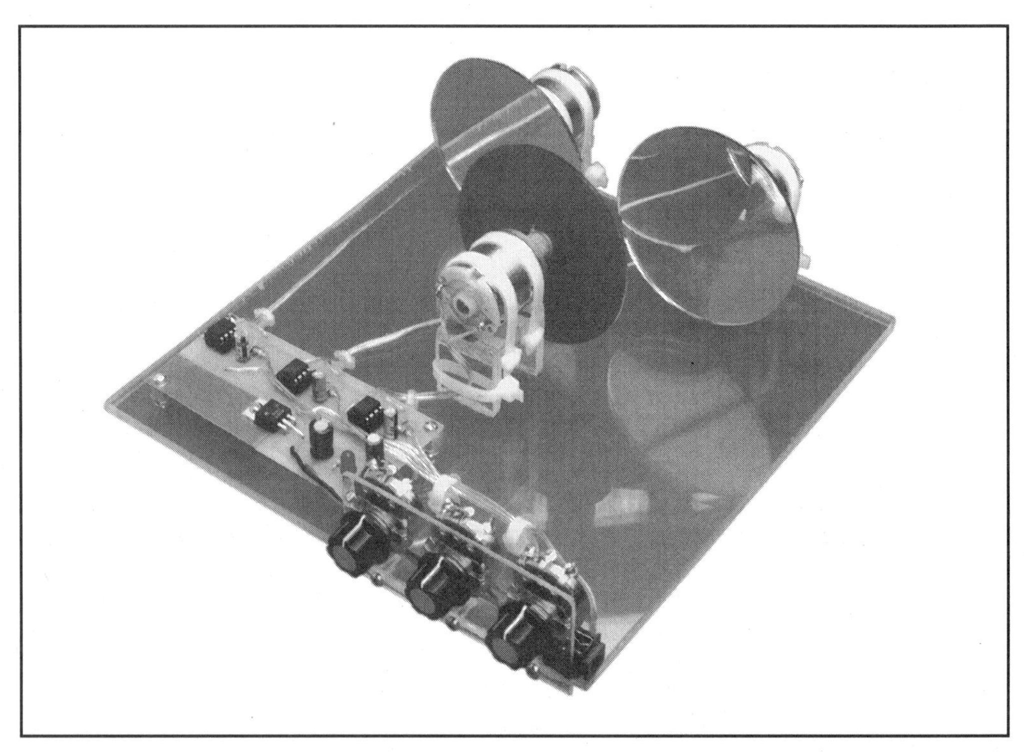

Photo 6-1. The completed laser light show. This unit is mounted on a sheet of clear plexiglass, but other materials will work just as well.

Now that we have our mirrors attached to our motors, these assemblies need to be secured to the main unit. For this, I used blocks of plexiglass tall enough to allow clearance for the spinning mirrors. These are glued to the base with PVC cement, and the motors are fastened with "cable tie" straps.

Moving right along, the motor control board is temporarily affixed to a reserved area of the base. It is best not to bolt the board in place just yet, as leads for the motors still need to be installed. In the prototype, the potentiometers are mounted on a panel made from a thin piece of plexiglass, which is attached to the edge of the base along the control board side.

Last, using hookup wire, connect the output of each oscillator to its corresponding motor. Now you can secure the control board. If you decide on separate units, I recommend a single cable to attach the two sections. By the way, be sure the oscillator output goes to the positive motor terminal or the motor will turn backwards (counterclockwise). I don't know, maybe you want it to turn backwards.

Before installing the three LM555 ICs, apply the external power and run a quick voltage check. The pin 8 and pin 4 locations on each of the sockets should read around 12 volts against the pin 1 (ground) location. If all is well, disconnect the power and plug in the 555s. The system is ready to test.

TESTING AND OPERATION

Testing is a simple matter of reapplying power and checking to see if the motors work the way they are suppose to. Start with all three potentiometers in the full counterclockwise position. With the external DC present, the motors will begin to turn as you advance the control pots, and gain speed as you approach the full clockwise position.

If one or more of the motors function in an opposite fashion, reverse the lines from the oscillator board to the potentiometer. This should correct the problem. Also, since the mirrors are slightly off perfect alignment, the motor speed should be easy to observe.

Once you have established that the motor system is cookin', it is now time to try introducing a laser beam. The helium-neon laser you built in Chapter 2 provides an exemplary source for this beam.

Photo 6-2. Closeup view of the motor/mirror section of the laser light show. Note the angles on the mirrors.

With the motors still, align the beam so that it reflects off all three mirrors and out the opposite side. Now comes the fun! Start the motors turning and watch the patterns from your new laser light show appear on the wall (or a screen).

You will notice that as you adjust the potentiometers, the design will change, sometimes quite dramatically. And, it is hard to duplicate a pattern. Now, this may be good news or bad news, depending on your perspective, but it does provide for a high level of versatility concerning application of this device.

The variation of lissajous designs affords marvelously entertaining effects, but it is difficult to reproduce a particular pattern unless you make careful notes regarding your control settings. If you anticipate the need to duplicate certain designs, I recommend you devise a calibration method for returning the potentiometers to the positions that generated the original pattern.

Also, combining this system with what you learned in Chapter 5 allows for even more variation in the emitted designs. By modulating the beam in a noticeable fashion, one additional element is entered into the overall equation. And, this can really make a vast difference in the final result.

CONCLUSION

So, what do you think? A nifty gadget, eh? Well, wait until you get yours built and working. This thing is nothing short of just plain fun! I have had mine for over six years, and it continues to garner interest from others and fascinate me.

Now, I know some folks are more easily entertained than others, but I do predict that you, too, will enjoy the heck out of this device. Trust me! It's a HOOT! I mean, more fun than a day at the beach, or a barrel of monkeys, or a trip to the mall (?) or...well, you get the idea.

On that note, now is the time for all good men to come to the aid...uh, that's the wrong speech. Now is the time for everyone to get their materials together and build the laser light show. Yeah, that's better. OK, enough nonsense!

On a more serious side, I doubt you will regret building this project. It is simple, inexpensive and very functional. And, at the risk of frightening someone, it's educational as well. Wrap all this up with a pretty bow, and you have one super endeavor.

CHAPTER 7

A LIGHT METER FOR LASERS

One of the nice things about surplus (and/or used) lasers is their price, but does their output really match their rating? For that matter, how about new devices? Since experimenters in this field often work with red wavelengths, and red is very difficult for the eye to evaluate, it is nice to have a method to accurately rate a laser's power.

That is where this project comes into play. Here we will build a simple light meter that reliably measures the beam intensity and displays the results digitally. For me at least, *digital* always makes reading test equipment far more enjoyable.

Once calibrated, this unit will allow you to gauge all your laser systems in terms of actual output as opposed to rated output. Hence, you will know precisely the amount of light you have. While that might sound more like a convenience than anything else, when it comes to such things as holography and scientific measurements, this information is very important.

With all that said, let's take a look at our nifty-thrifty, handy-dandy laser light meter. It is simple to construct, easy to use and relatively inexpensive. Besides, it uses a chip that is really nice to work with, the Intersil/Harris ICL7106 analog-to-digital converter.

THEORY

I have been plugging 7106/7107s into my prototypes for as long as I can remember (Of course, sometimes that's only as far back as last week.), and they have never failed me. Anytime you want to measure a direct current (DC) voltage, or any value that can be converted to a DC voltage, you'll have to search far and wide to find a better method than these gems.

So, since a 7106 is employed in this circuit, you know I'm gonna like it. But, rather than just taking my word, let me tell you a little about these chips. Each houses, in a single 40-pin dual in-line pin (DIP) package, an analog-to-digital converter, reference, clock and digital decoder/driver circuits.

The 7106 also includes all that is necessary to drive an *instrument size* (0.5 inch) liquid-crystal display (LCD), while the 7107 illuminates equally sized light-emitting diode (LED) elements. Thus, you have two choices when designing a project.

The LEDs are a little easier to use, but they consume far more power than their LCD cousins. So, if battery power is in the equation, as it is with this project, the 7106 is your boy! It will also operate off a standard 9-volt battery, as opposed to a split (+/-) supply, requisite for the 7107.

Like many analog-to-digital converters, these employ a *dual slope* arrangement where the reference voltage (first slope) is measured against the time it takes for a capacitor to discharge (second slope). The capacitor is initially charged by the incoming signal, making it representative of said signal. Hence, the discharge time provides an accurate indication of the input.

Once the measurement has been taken, it is fed to the decoder/driver goodies, and voila!...you have that measurement in big, easy-to-read numbers. As I said, that's my kind of meter. Except with communications equipment, I'll chose a digital display over an analog meter anytime!

The second half of the circuit provides light detection. In this section, operational amplifier (op-amp) U2 is configured as a noninverting comparator. Referring to *Figure 7-1*, note that potentiometer R7 is connected across the power source, and the op-amp's inverting input (pin 2) is tied to the *whipper* (center connection). This generates the circuit's *reference voltage*. Photoresistor CdS1 is wired in series with resistor R6 and this network is also connected across the power. U2's noninverting input (pin 3) comes off the CdS1/R6 junction and acts as the *voltage in* (Vin) half of the comparator.

Since the photoresistor changes resistance when light strikes its surface, shining the laser beam on CdS1 will raise the Vin, breaching the reference threshold and triggering the op-amp. As the beam's intensity increases, so does the amp's output, and this allows the digital voltmeter to measure and display a voltage proportional to the magnitude of the light.

The last part of the device, resistors R8 and R9, comprise a voltage divider that allows the 7106 to display readings as high as 20 volts. However, the battery voltage is only 9 volts, so that is as high as the meter will read. More will be said about this in the calibration section.

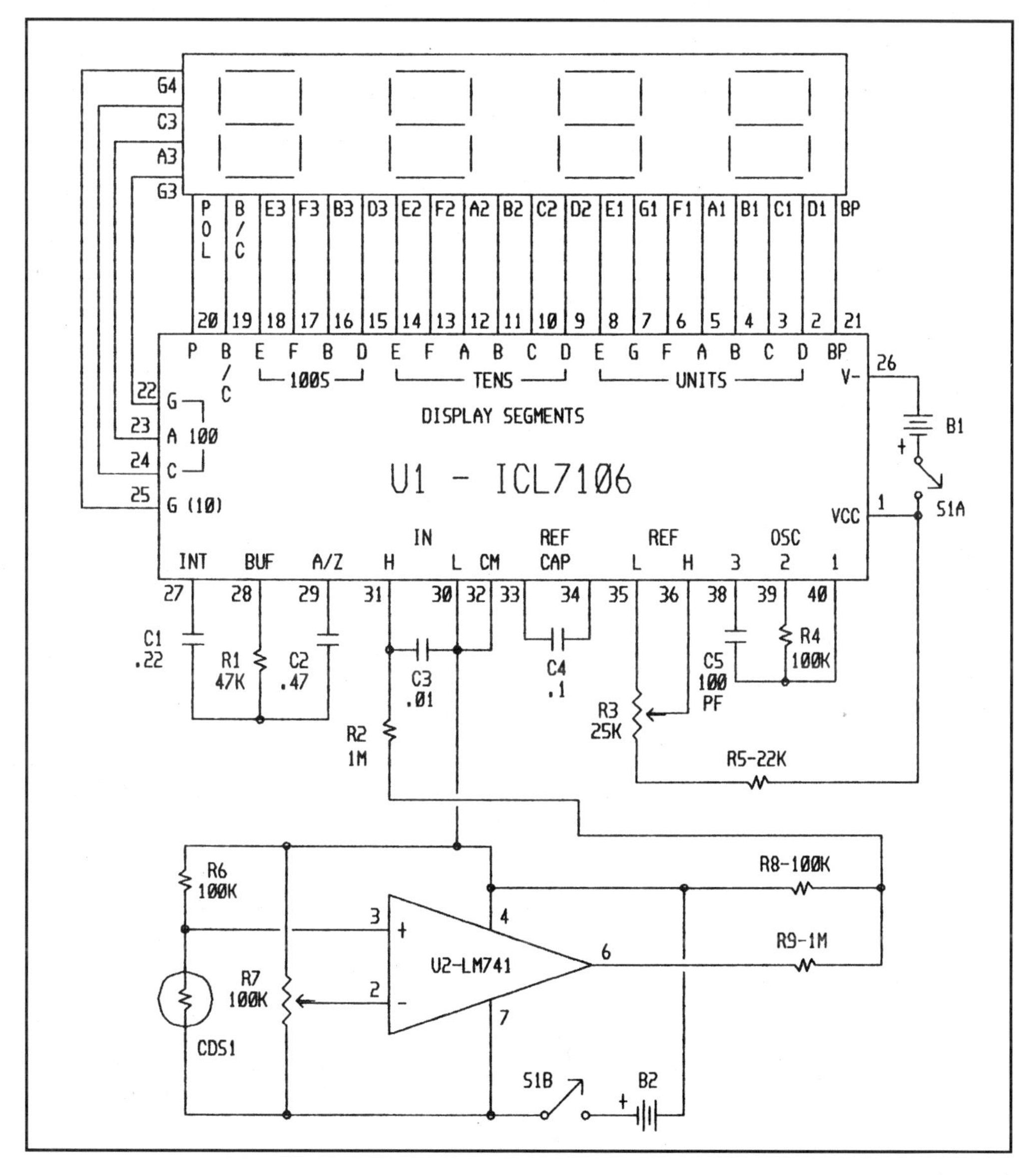

Figure 7-1. Laser light meter schematic.

You will also note from *Figure 7-1* that two separate 9-volt batteries are needed. This is due to the fact that the 7106 can not read its own power source. If we tried to use the same battery to power both sections, the AD converter would be attempting to do just that. A conflict would occur, and the results would not be good at all; hence the necessity for two 9-volt cells. Also note that switch S1 handles "power on" for both sections (batteries).

When you put all this neat stuff together, you have a very sensitive light meter that will read the beams from your lasers. OK, enough about that. Let us move on to the construction phase. After all, this thing isn't going to do anything until we get it built. Right? Right!

CONSTRUCTION

This project can be built in two different ways. You could start from scratch and assemble each section from parts, or you could fabricate the op-amp/voltage divider on a printed-circuit board (PCB), then add a DC voltmeter module.

The module is factory constructed and ready for use, which does accelerate the process. Hence, this is the approach I selected; you know, for expedience. Yeah! I hear you! You're saying, "He's just lazy." Well, not really. I have built a few 7106/7107 voltmeters in my time, but if you have your heart set on doing so yourself, the schematic is included in *Figure 7-1*.

If you want to follow MY program, however, let's start with *Figure 7-2*. It is the PCB foil pattern for the op-amp section of this marvel of electronic ingenuity. I have also included *Figure 7-3*, the parts placement diagram, for your convenience.

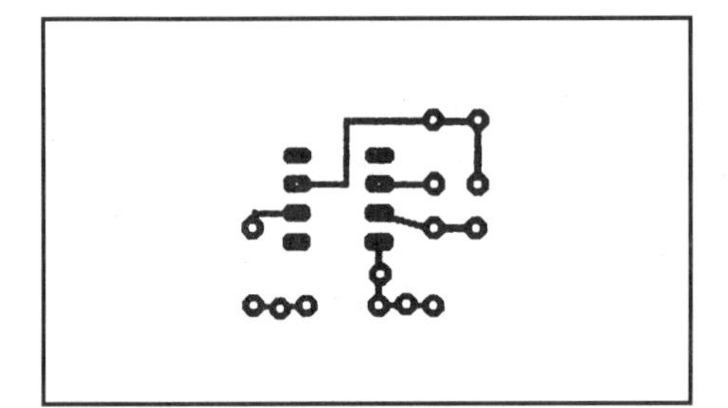

Figure 7-2. Laser light meter PCB foil pattern.

With the PCB in hand, first install the IC socket, then resistors R6, R8 and R9. As you can see, this isn't one of your more complicated designs. In fact, if you don't want to make a printed-circuit board, it can easily be done on perf-board with point-to-point wiring. I just happen to like PCBs—but then, I guess you know that by now.

LASER LIGHT METER PARTS LIST

SEMICONDUCTORS

U1	Harris/Intersil ICL7106 A/D Converter
U2	LM741 Operational Amplifier
DSPLY1	3-1/2 Digit LCD Display (Jameco)
MOD1	LCD Digital Voltmeter Module (Jameco, JDR, BG Micro)

RESISTORS

R1	47,000 Ohm 1/4 Watt Resistor
R2, 9	1,000,000 Ohm 1/4 Watt Resistors
R3	25,000 Ohm PCB Potentiometer
R4, 6, 8	100,000 Ohm 1/4 Watt Resistors
R5	22,000 Ohm 1/4 Watt Resistor
R7	100,000 Ohm Panel Potentiometer

CAPACITORS

C1	0.22 Microfarad Mylar Capacitor
C2	0.47 Microfarad Mylar Capacitor
C3	0.01 Microfarad Mylar Capacitor
C4	0.1 Microfarad Mylar Capacitor
C5	100 Picofarad Ceramic Capacitor

OTHER COMPONENTS

CDS1	Cadmium Sulfide Photoresistor (Radio Shack)
S1	DPST Slide Switch (DPDT will also work)
B1, 2	9-Volt Alkaline Batteries

MISCELLANEOUS

PCB Materials, Solder, Hookup Wire, Hardware, Case, Battery Clips, 1/2 Inch PVC Pipe, IC Sockets, Knob, Paint, Dry Transfer Material, Etc.

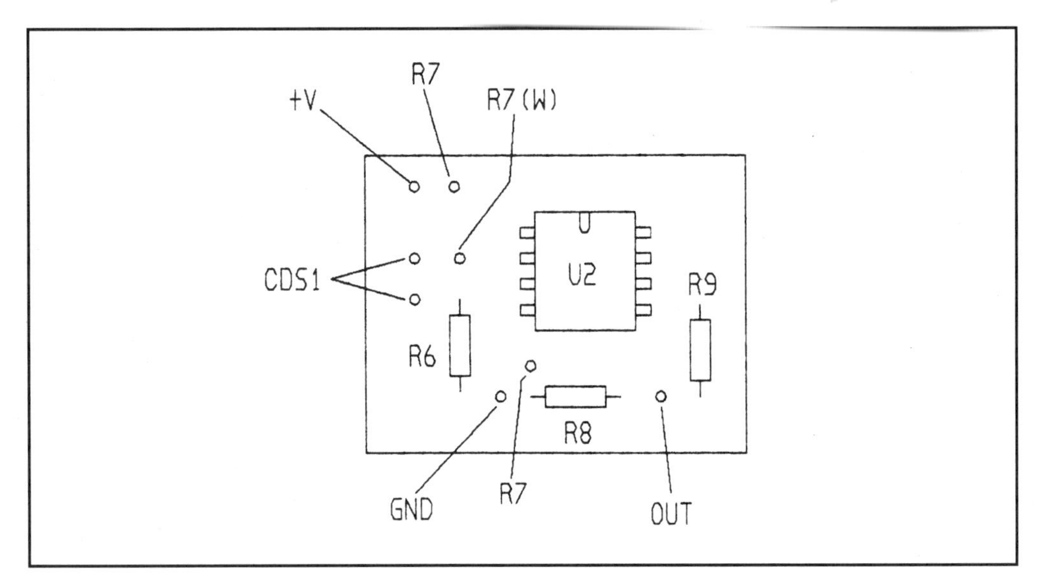

Figure 7-3. Laser light meter parts placement.

With the PCB in hand, first install the IC socket, then resistors R6, R8 and R9. As you can see, this isn't one of your more complicated designs. In fact, if you don't want to make a printed-circuit board, it can easily be done on perf-board with point-to-point wiring. I just happen to like PCBs—but then, I guess you know that by now.

The next step is to solder appropriate lengths of hookup wire to the board for potentiometer R7, photoresistor CDS1, the 7106 module and switch S1. Since nothing has to go through the panel before soldering, you can connect the other ends of the hookup wire to the various components.

With all that done, set the electronics to one side, and let's move on to the case. Alright! Alright! If you must, connect the batteries and give the little devil a quick test. You should see a reading on the display that will change as you adjust the potentiometer and/or light level on CDS1. However, we will be covering this procedure in greater detail later.

For the case, just about any project box large enough to handle everything will do right nicely. Don't forget you need room for two 9-volt batteries. With the prototype (see *Photo 7-1*), I employed a standard plastic project box of dimensions

4-3/4 inches (121 mm) x 2-9/16 inches (65 mm) x 1-9/16 inches (40 mm). This easily accommodated the batteries, PCB and module.

In the top, I cut a hole for the module, which is attached with plastic standoffs. This hole needs to be large enough to allow viewing of the display, yet small enough to provide a mounting area. However, if a different type of module is employed, the opening will have to be cut to indulge that unit.

Another opening, in the side of the case, is made for the switch. I used a small slide switch, and fashioned the opening by first drilling two holes roughly the size of the slider, then filing them out to conform to the switch's size and shape.

Finally, a last hole has to be drilled for the *range* potentiometer. This ain't rocket science, and merely requires boring an aperture with the right size drill bit. If the hole allows the shaft to fit through, and isn't too big, you've done it right. And that's all I'm going to say about that.

Once the lid is completed, give it a coat of light yellow paint. This color allows the black dry transfer lettering to be easily read, which is nice, as that way you will know what you are doing. Anyway, after the paint dries, label the various functions.

The top is then given a LIGHT coat of clear enamel to protect the fragile lettering. I stress *light*, as much of the dry transfer material will melt if you get this coat too heavy. Believe me, that really makes a mess of things.

While the protective coat is drying, you will want to drill a hole in the bottom section large enough for the CdS photoresistor arrangement. I put the sensor at the top (above the meter), and used an old dome-style lamp holder for the sensor.

The CdS cell is glued to the base of the lamp socket with epoxy, and the original contacts can be used for connection to the circuit. Something nice about this approach is that various domes can be employed. A white one does a great job of diffusing the beam, while red or green domes are handy for beams of those colors. OK, back to the main gig. With the paint dry, you are ready to install the panel parts. When that's done, it is time to test the unit. If you have built the voltmeter from scratch, you will also need to calibrate it.

Photo 7-1. The completed laser light meter. The red dome on top allows the meter to better detect the 632.8 nanometer wavelength of helium-neon lasers.

CALIBRATION AND TESTING

For the purpose of calibrating this device, you will need a laser of known power output. Probably the best choice here is either a new 1 milliwatt helium-neon tube, or a laser pointer. Usually, the new tubes are precisely on target in terms of output, and so are the pointers.

The pointers are, of course, easier to find and will usually be in the 3 to 5 milliwatt range. So this may well be your best bet. Another source of highly accurate lasers is the physics department of your local university or college. Their lasers need to have fairly precise tolerances.

If you are courteous, they will probably allow you to come over and calibrate your meter. But, be polite! Although I don't have to tell you that, do I?

Anyway, with a suitable laser light source in hand, shine the beam on the photoresistor (CDS1) and adjust potentiometer R7 until the LCD display matches the laser's power rating. If you have more than one reliable source available, try each to see if the meter reads the beam correctly. This is especially helpful if they are of different power outputs.

One slight drawback to this configuration is that it will only read powers up to 9 milliwatts. This is because the measurement section of the system (U2 and friends) is powered by a 9-volt battery. Thus, the maximum value that can be displayed will top out at "9," or 9 milliwatts. However, you probably won't be working with lasers even that large, at least not at first, so this shouldn't be a problem.

Normally, once calibrated, this meter will be right on the mark until you have to change batteries. At that time, a recalibration may well be in order. And, that is why I included R7 on the front panel. It makes this process more convenient, but if desired, the potentiometer could be mounted on the PCB—with slight modification to the board, of course.

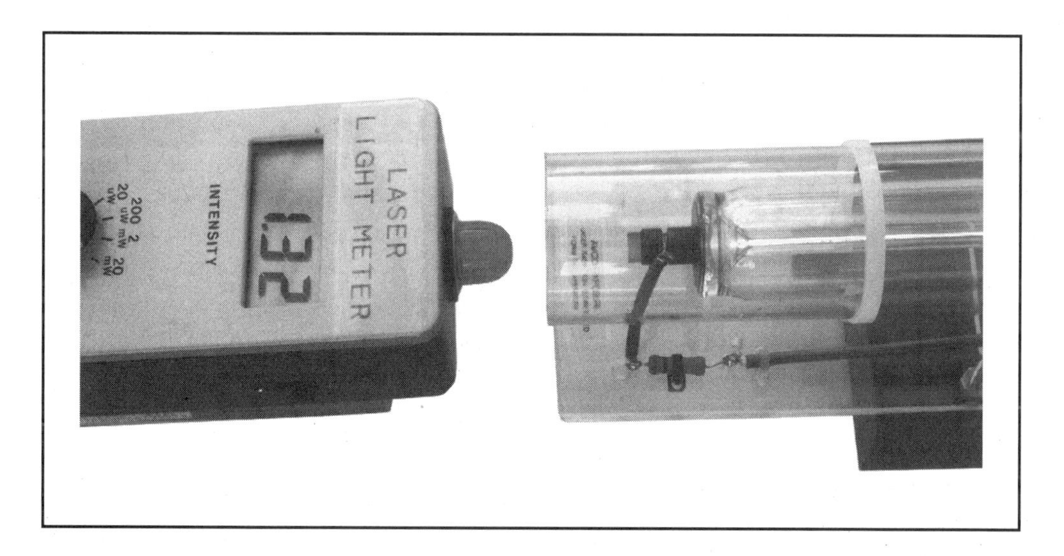

Photo 7-2. The completed laser light meter in operation. Note the 1 milliwatt laser is actually delivering 1.32 milliwatts of power.

If you elected to build the "whole enchilada," the A/D converter, or voltmeter, will have to be calibrated. This is a fairly simple procedure. Since the basic voltmeter is configured for 2 volts full-scale, it will need a reference voltage of 1 volt between pins 35 and 36.

So, hook up a DC voltmeter to those pins (positive on pin 36) and adjust potentiometer R3 until you get a 1-volt reading. That will set the optimum reference voltage for the 7106.

As for testing, try aiming a laser of unknown output power at CDS1. The meter will give you a reading corresponding to the actual output of that laser. If you don't have access to an unknown laser, don't worry. It probably won't be long before you do, and besides, as you will find in Chapter 8, this meter has other purposes.

OPERATION

While helium-neon tubes perform admirably for several thousand hours (on average), their output level will diminish gradually with age. This device will allow you to periodically check your laser's power. That way, you will always be aware of the true output. And, this can be quite important when you get into areas such as spectroscopy, holography or materials evaluation.

Naturally, the process is the same as for calibration and testing; that is, point the laser's beam at the CdS cell and note the display reading. Oh, don't forget to turn on the power. You know, both the meter and laser. Alright, I take that back. I'm sure you will fire up both of them before starting.

Incidentally, this meter will also read regular light. However, it will require a different calibration for that task. So, if you need a light meter for, say, photography, this could easily fit the bill. At least, it's a great place to start.

CONCLUSION

Well, that's about it. You are now the proud owner of a "Laser Light Meter" that you built with your own two/three hands. And, it will come in handy! Trust me!

Holography is the subject of the next chapter, and that is a field that definitely will find a purpose for this device. Keeping track of light intensity levels within many holographic setups is next to essential. But more will be said about that in Chapter 8.

Also, it is always at least "nice" to know how much power your laser(s) is putting out, not to mention being able to check those new tubes and diodes you order. In reality, no laser is so precisely made that it emits the exact power advertised.

But, with your trusty light meter at your side, they "ain't a gonna fool you no more!" You'll note that in addition to crossing all my "T"s and dotting all my "I"s, I pride myself in always using proper grammar!

OoooooooK! All kidding aside, this will prove to be one handy tool for your laser lab. It is easy and inexpensive to build, and fascinating to use. Once built, I think you'll be astounded by the results you get with this unit. And, as revealed by this meter, the amazing way laser light responds to various influences.

CHAPTER 8

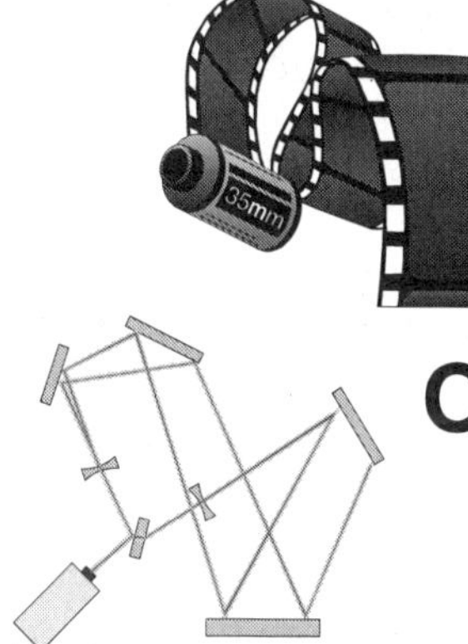

HOLOGRAPHY, OR 3-DIMENSIONAL PHOTOGRAPHY

Holography is the art of taking an object and making a three-dimensional (3-D) picture of it. These are called *holograms*, and I'm not talking about those post-cards with a prism-like overlay, or the stereoscopic shots that require a viewer. With holograms, a true "third dimension" is achieved. True in the sense that as you rotate or move around a hologram, all of the object can be viewed.

Born from a novel marriage of science and art, holography rests on a sturdy foundation of optical physics. However, it was the artistic community that pulled it up by its bootstraps. Through small schools, museums and studios in San Francisco and New York, not to mention the dedication of a handful of industrious individuals, the world was introduced to holography in the early and mid 1970s.

Some of these fascinating photographs (transmission holograms) do require a laser to observe them, but that is not a problem, as you can use the same laser employed in shooting the hologram. In fact, it is recommended that you use that same laser. That insures the highest-quality reproduction.

A second form, the *reflection hologram*, can be viewed in normal white light, but the shooting arrangement doesn't result in as much depth of field as with transmission shots. In this chapter, we will learn how to make both types, which hopefully will also furnish a good infrastructure for further work in this field.

A well prepared hologram is something to behold! You have to see it to truly appreciate it, and if the first hologram you view is one you made yourself, the impact is intensified. It really makes an already exhilarating event even more stimulating.

I will, at the outset, confess that this area of laser exploration is a tad more complicated than some other arenas. This is not to say that it's difficult, just a little more involved. But trust me, it's worth it!

So, let's get our feet wet! Once you have the basics in place, the rest comes easily. And, it is a lot of fun! Of course, I find ALL this stuff fun, but holography tends to fall into an *entertainment* category all its own.

THE BASICS

In terms of this realm, the basics are, well...pretty basic. You will need a selection of both positive and negative lenses (see Chapter 4), some other optical elements, a positioning table and naturally, a laser. You will also need a room, or other area, that can be made completely dark.

The most complex part of all this is the table, so let me cover that subject first. Again, complex may be a poor choice of words, as this isn't brain surgery. It is simply the primary step in putting you on the trail to holography.

Check out *Figure 8-1*, and you will see a typical *positioning table* for hologram production. This one runs about 6 feet by 4 feet and incorporates all the elements required. The most important of those elements is stability; that is, you must have a table that dampens out as much vibration as possible.

The exposures used in shooting a hologram are relatively long (depending on the laser's power, they can be an hour or more), and your biggest enemy is vibration. ANY shaking, no matter how subtle, will effect the sharpness of your shot. And, if enough *rockin' and rollin'* occurs, the hologram will be ruined. Obviously, you don't want that to occur.

OK, the "main bed" is probably as good a place to start as any. This will determine the overall size of the table, which should be of adequate dimensions to handle any of the setups you anticipate doing. For this text, I recommend a bed 4 feet (1-1/3 meters) wide by 6 feet (2 meters) long and between 3 to 4 inches (75 to 100 millimeters) deep.

To date, these measurements have always proved more than adequate space to set up virtually all but the most complex optical configurations. And, this size also provides enough weight to admirably manage most vibrations encountered.

The bed and frame can be made from a variety of different materials, but I have found wood to be the best. It is easy to work with and usually less expensive than metal or plastic stock. And, normally, it is far simpler to obtain.

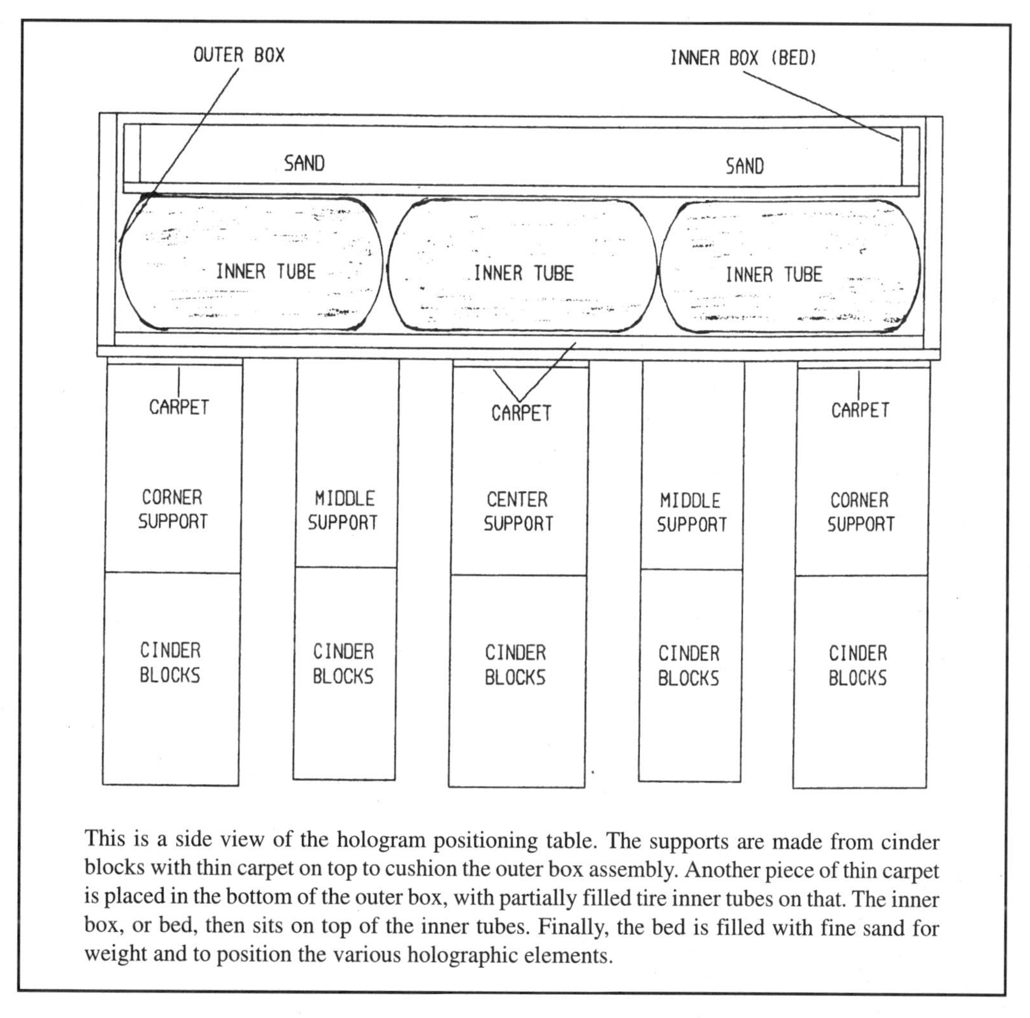

This is a side view of the hologram positioning table. The supports are made from cinder blocks with thin carpet on top to cushion the outer box assembly. Another piece of thin carpet is placed in the bottom of the outer box, with partially filled tire inner tubes on that. The inner box, or bed, then sits on top of the inner tubes. Finally, the bed is filled with fine sand for weight and to position the various holographic elements.

Figure 8-1. Diagram of hologram positioning table.

So, using wood, the bed frame is cut from 1 x 4 inch boards, and the bottom from 1/2 inch plywood. The bottom is fastened to the frame with screws and wood cement. Be sure the seam between the two is tight, as the bed will later be filled with sand, and you don't want sand running out all over the place.

Next comes the outer box that will hold the bed. This needs to be just slightly larger than the outside dimensions of the bed—in other words, the bed should not touch the sides of the outer box. You will be placing automobile tire inner tubes and carpet in the bottom of this structure, so it has to be deep enough to handle that and the bed. I used 1 x 12 inch boards for the sides and 1/2 inch plywood for the base, again secured with screws and cement.

Now comes a decision: where to place your holography table? As stated before, the area has to be completely dark when shots are being made. And, once in place, you will not want to be moving it around. This sucker's heavy! Hence, your location should be one that's at least somewhat out of the way.

In the past, a corner of the basement, that can be light secured, has proven a perfect spot to put the table. That is, of course, when you have a basement. If you don't, a large closet or spare room usually serves quite well. However, if at all possible, stay on the ground floor, as the structure's foundation will add stability.

With your location selected, the *base* for the outer box is next. I have yet to find a better material for this than cinder blocks. You know, those large, heavy, gray-colored cement bricks with holes in the center. They are tremendous at absorbing vibration, more than sturdy enough, and CHEAP!

With this size table, you will need 3 foot (1 meter) high legs at each corner and halfway down the sides. Also, on each end of the table, a leg should be placed down the center roughly halfway between the corner legs and middle supports.

Be sure this arrangement is level before placing the outer box on top. An uneven base tends to amplify any strong vibrations that effect your area. Also, it makes placing the various elements, and keeping them in place, harder.

Now, put sections of relatively thin carpet on top of each leg and set the box on those. Lay another section of carpet, cut to size, in the bottom of the outer box, then place the inner tubes on the carpet. The tubes should about 2/3 full; that is, firm but not hard.

With all that positioned, the bed, or inside box, is laid on top of the inner tubes. The next step is to fill the bed to near capacity (1/2 inch from the top) with white sand. This is the stuff you buy at home and garden supplies for sand boxes, concrete, et cetera. It should be fairly fine and as clean as possible. And, you are going to need quite a bit of it (probably 500 to 600 pounds).

The sand serves two purposes: First, it contributes a majority of the system weight, providing the needed stability; and second, it furnishes a means of holding the various components in place. The lenses, mirrors, film holder, beam splitters and other devices can be positioned easily in the soft, but dense, sand.

When you have returned from the hospital following the operation for that hernia you developed by lifting all the sand, you're ready to start making holograms (Just kidding about the hernia!). However, some holographers like to mark the edges of the bed with letters and/or numbers. This provides *benchmarks* they can record, which are especially useful when duplicating a setup. If this idea *grabs* you, now is the time to do it.

And, that is how you go about setting up your positioning table. The next phase is to shoot a simple hologram. They come in two flavors, *reflection* and *transmission.* (Actually, there are some far more complex versions—for example, *image plane hologram*—but we're not going to get into those) So, let's move on and explore each of these techniques.

THE TRANSMISSION HOLOGRAM

As stated above, transmission holograms must be viewed with laser light. This approach provides much greater depth of field (depth sharpness) than its cousin the reflection hologram and is the perfect place to start.

Well, at least that is my feeling. I have encountered some rather stubborn opposition to this in the past. I don't know! I just can't understand how some people just don't see things my way. Can you?

Anyway, the transmission hologram setup is depicted in *Figure 8-2*. And, as can be seen, it is a very simple arrangement. The laser beam is spread with a *negative* lens so that it covers both the object being photographed and the film (holder). The position of the lens is maintained by sinking the lens holder into the sand.

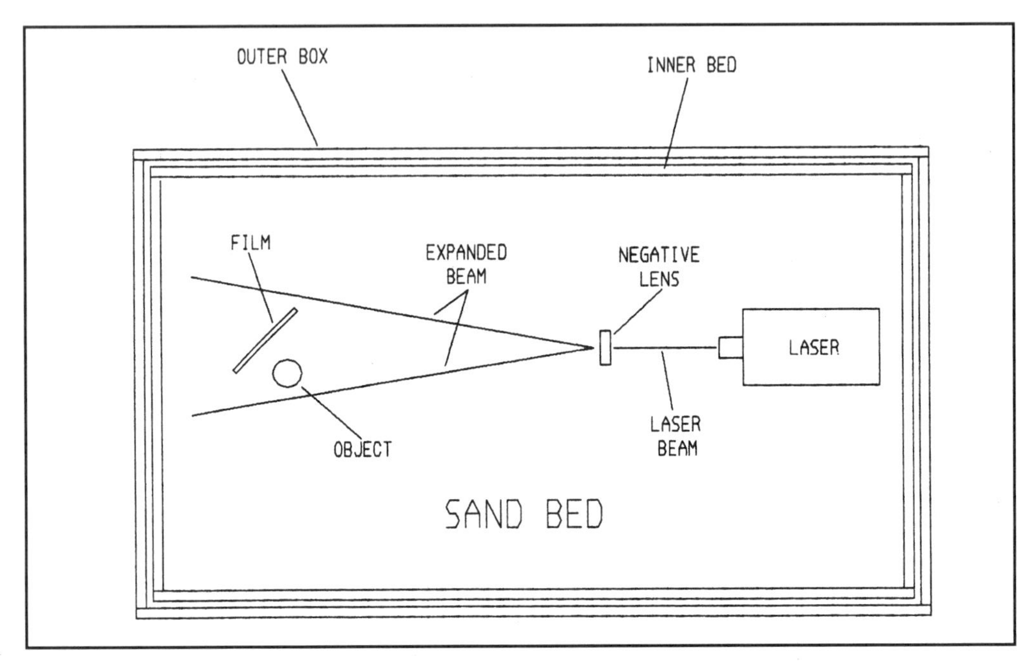

Figure 8-2. Simple *transmission* hologram setup (overhead view).

You will want to adjust this lens so the light just illuminates both the film and object. Light that extends past either is called *spillover* and is wasted light. This results in even longer exposures, so try to just cover the object and film. Naturally, that is often easier said than done, and you don't have to overdo the project. If you can keep the spread close to the above recommendation, that will be good enough for government work. You are going to loose some light. There simply isn't any way around it.

Both Kodak and Agfa-Gevaert produce film for holography, and I personally like the 4 x 5 inch sheet film. This stuff does require a 4 x 5 film holder, but in a way, that is good. These holders have *slides* (one on each side) that protect the film from light, and this makes working with the film much easier.

Take one of these holders and install it in the sand at roughly the angle pictured in *Figure 8-2*, adjusting it until the object lines up with the center of the holder. That will insure you photograph all of the object. Depending on its size, the subject may have to be elevated slightly by placing it on a small mound of sand.

Once you have everything in place, it's time to shoot the image. The holder will have to be removed from the setup and loaded with film by removing the slide(s) and slipping a sheet of film in from the top. This, of course, requires complete darkness and don't forget to replace the slide(s). With the holder loaded, turn on the lights, and return it to its proper position. Now, you're ready.

The first step is to turn on the laser. The second step is go take a nap, watch TV or eat dinner, lunch, whatever. You can do any of these things because the laser needs to be operational for at least an hour before you shoot the hologram. That allows it to stabilize.

When the laser is ready, so are you. Kill the lights, well...not really! Just turn them off. Place a piece of black cardboard in the path of the laser, to act as a shutter, and remove the slide from the film holder. Now, remove the shutter and begin timing the exposure. *Table 8-1* provides some exposure guidelines for various laser powers and setups.

This table is provided as a starting point for various laser power levels and hologram configurations. Since each laser is different, all optics are not the same and setups can vary, this table can only be a general guide. In many cases, you will have to experimentwith your laser and optical layout to get a better handle on what the exposures should actually be. Eventually, after some practice, you will learn how your system functions and be able to shoot and process nearly perfect holograms every time.

LASER POWER	HOLOGRAM TYPE	TIME
1 MIILIWATT	TRANSMISSION	3-5 MINUTES
3 MILLIWATTS	TRANSMISSION	2-4 MINUTES
5 MILLIWATTS	TRANSMISSION	1-2 MIUNTES
1 MILLIWATT	REFLECTION	3-5 MINUTES
3 MILLIWATTS	REFLECTION	2-4 MINUTES
5 MILLIWATTS	REFLECTION	1-2 MINUTES
1 MILLIWATT	SPLIT-BEAM	8-10 MINUTES
3 MILLIWATTS	SPLIT-BEAM	4-7 MINUTES
5 MILLIWATTS	SPLIT-BEAM	2-5 MINUTES

Table 8-1. General hologram exposure times.

At the end of the shot, replace the shutter then the film holder slide. The hologram has been exposed. The film still has to be processed, but we will cover that in a little while.

THE REFLECTION HOLOGRAM

Here we have a technique that allows the finished product to be viewed in white light as opposed to laser light. Referring to *Figure 8-3*, you will see this approach is very similar to the transmission hologram strategy. The major difference is the film is positioned in front of the subject instead of to one side.

Again, a negative lens is employed to spread the beam so that it will cover both the film holder and subject. However, with this arrangement, the object being photographed is behind the holder, so the spread should provide a little spillover to insure there is reflection off the object.

As with the transmission procedure, you will want to line up the object with the empty holder. This allows you to position your subject and be sure all of it is recorded.

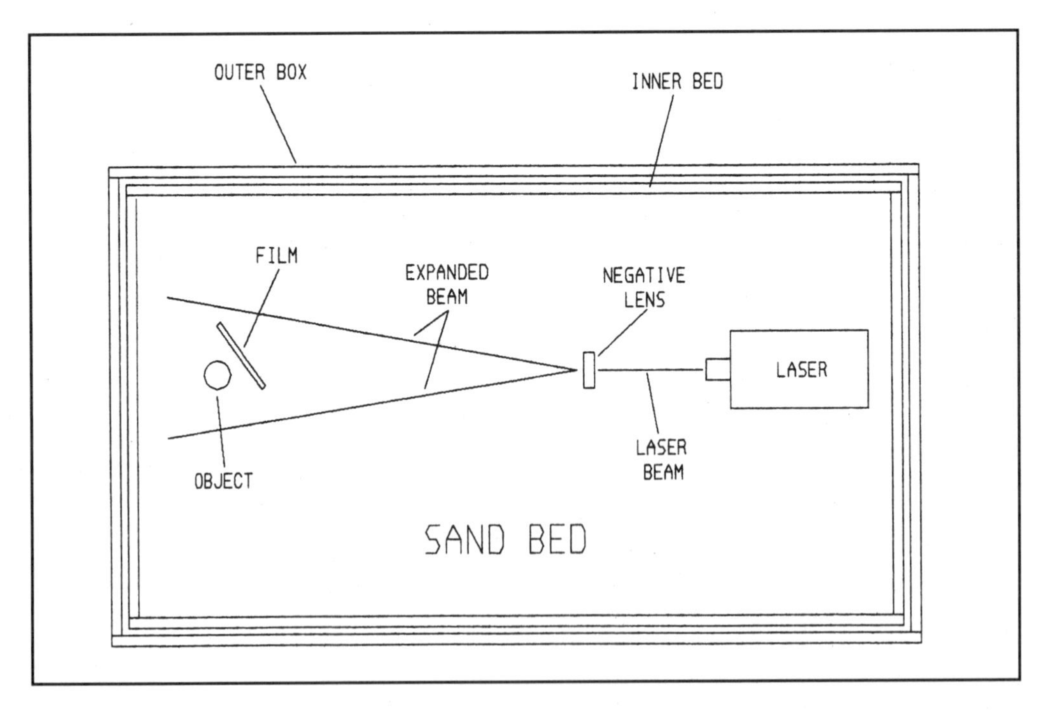

Figure 8-3. Simple *reflection* hologram setup (overhead view).

The same shutter arrangement is used here to block the beam while the film holder slide is being removed (before the exposure) and replaced (after the exposure). Also, the laser must again warm up for at least an hour before shooting.

And, as before, both the film loading and exposure must be done in total darkness. These films are *panchromatic* in nature, which means they are susceptible to all visible light wavelengths. Hence, any stray light will result in film *fogging* (extraneous exposure), and that will damage, or even destroy, your hologram.

Alright! If you can shoot a transmission hologram, the reflection type shouldn't give you any trouble. One quick word of caution, though. If you plan on shooting both types, be sure to label which is which. This is because you can use a different processing procedure for each type.

OK, you're doing great! However, these holograms are not the very best that can be produced. Since we are using a single beam, some degree of *shadowing* will occur. And, in terms of quality, this leaves something to be desired.

The reason for the problem is that both the reference beam and object beam are coming from a single source. However, it isn't all that hard to correct this dilemma. You will have to surrender some of the simplicity of the single beam setup, but regarding quality, the resulting holograms will be far more pleasing to the eye.

So let's again move on to a different approach—the *split-beam* hologram. I think you will like this approach, and it will give you a chance to use a few more of those optics you have "squirreled" away.

THE SPLIT-BEAM HOLOGRAM

As its name implies, the split-beam method involves using a single laser beam and splitting it up into two or more beams. The purpose here is to provide separate object and reference beams, and that helps eliminate the inherent shadows of single-beam holograms. In some ways, you have a compromise between a transmission and a reflection hologram.

This approach does require more optical elements, which tend to absorb some of the light, but increasing the exposure time will compensate for this loss. All in all, you are probably going to be happier with the results from a split-beam shot as opposed to the single-beam technique.

The path distances for the various beams should be as close to equal as possible. That is, the light span from the laser to the film (the reference beam) should be about the same as the object beam's span from laser to subject.

Referring to *Figure 8-4*, you will see that split-beam holography employs lenses, mirrors and beam splitters. The outcome is a dual-beam strategy where one set of light shafts bathe the object (subject), while the second (reference) beam illuminates the film. In this fashion, both the film and the object are well lighted, and most of the offensive shadows fill in.

Incidentally, mirrors used for holography must be of the *front-surface* variety (see Chapter 4). This is due to the fact that rear-surface mirrors produce a *double reflection*. The beam will reflect off both the front surface and the rear silver coating, and this causes problems regarding proper hologram production.

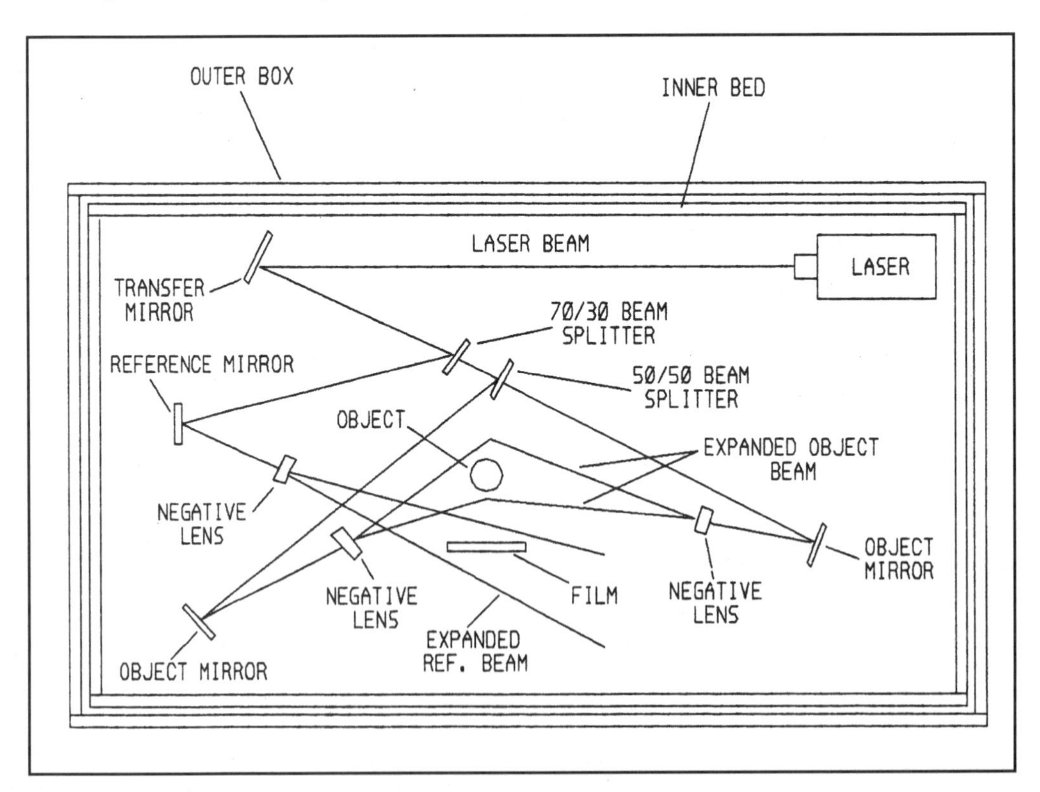

Figure 8-4. A *split-beam* hologram setup (overhead view).

Ready? Let's go! The laser is first reflected off the transfer mirror, then split with a 30/70 beam splitter. The 70-percent beam is then reflected off another mirror (the reference mirror), through a negative (spreading) lens and onto the film holder.

The 30-percent beam is fed to a second beam splitter with a 50/50 ratio. Each side of this split proceeds to an object mirror, through another negative lens and onto the subject (object). This produces object lighting from two positions at roughly a 120 degree angle.

Both the 30/70 beam ratio and the 120 degree spread are used to suppress noise. With holograms, noise will appear in the scene as *fog*. But, by using a reasonably large angle between the object beams and keeping the reference beam much brighter than the object beams, an impressive amount of this fog can be eliminated.

By the way, the relative strengths of the two beams can be measured with the light meter we built in Chapter 7. Take the readings at the film holder and object locations. With a typical hologram, you should see somewhere in the neighborhood of a 4:1 reference/object beam intensity ratio.

So, it is obvious this approach is considerably more complex than using a single beam to light both the film and the object. And, split-beam setups can get even more so. Say, three object beams, or maybe four. This, of course, would include additional optics to generate the extra beams.

As with all holographic layouts, you want to confine the beams to just covering their target. Any spillover from the film or object costs you in terms of efficiency. Thus, choose and place the lenses so the light barely blankets each component.

Naturally, all this beam reflecting and beam splitting and beam spreading takes its toll on the laser's output. Hence, the higher the light output, the better. I have found that with split-beam arrangements, a 3 to 5 milliwatt laser is a good choice. Those power levels provide plenty of light to shoot this type of hologram and still retain fairly short exposures. That, of course, is nice, but certainly not essential. If your positioning table is solid and stable, vibration, even during long exposures, will not be a problem. And, the 1 milliwatt laser from Chapter 2 will do just fine.

Previously mentioned procedures still apply with the split-beam approach. That is, be sure to *sight* your subject against the film holder, and load and shoot the hologram in total darkness.

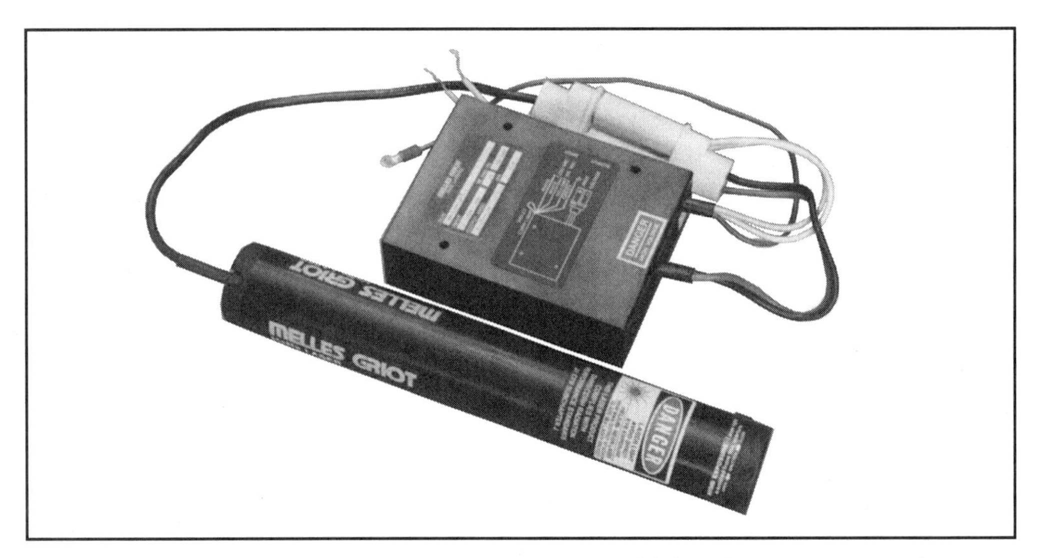

Photo 8-1. A 5 milliwatt helium-neon laser and high-voltage power supply. This one, by Melles Griot, is surprisingly small.

One additional technique, somewhat requisite for more elaborate setups, is *carding off*. Since stray light from the various optical elements tends to wander about, striking both the film and object, it is good practice to place black cardboard in its way. Carding off is the holographer's expression for doing just that.

With the object and the film holder in place, and the laser on, turn off the main light. Let your eyes adjust to the darkness, then observe any stray illumination coming from the likes of beam splitters, mirrors or other system elements. Now, place the black cardboard in such a way as to prevent the stray light from hitting either the object or film. Be sure, however, you do not inhibit any of the contrived lighting necessary for the hologram shot.

As for processing, split-beam holograms use the same scheme as reflection holograms. But this will be covered in the next section. So, stay tuned as it is coming right up!

HOLOGRAM PROCESSING

As has been previously mentioned, albeit only briefly, there are different processing methods for both transmission and reflection holograms. However, this only partially applies to Agfa-Gevaert film. With Kodak film, both types are processed (run) in the same fashion.

As you have probably gathered by now, Agfa-Gevaert and Kodak are the two principal suppliers when it comes to holographic film, plates and processing chemicals. Both companies produce excellent products for both red light and blue/green light holograms. But, since most of us use helium-neon lasers, the red light film and chemicals are what we need.

One eminently respected technique is called *pyrochrome* processing. This is a three-solution operation in which the first two solutions are mixed together, just before you start, to produce the developer. After development, the film is then *bleached* with the last solution, washed and dried.

With this process, the red helium-neon beam can be used to produce red, orange, yellow or green holograms by adding specific amounts of an additional chemical to the developer. Also, pyrochrome can process both reflection and transmission type holograms, furnishing high brightness and low noise. This one is a favorite among many holographers.

The three solutions are quite simple, and *Table 8-2* lists the chemicals needed for the pyrochrome process. Most photo supply shops worth their salt carry these chemicals, so you shouldn't have any trouble along those lines.

For white light holograms, Agfa has several different systems to choose from. These involve different types of developers, bleaches and fixers, and since they are constantly being improved, it is best to consult a photo supply dealer concerning what you will need.

Most Agfa schemes utilize a two-part developer, bleach and fixer, and just about all of these developers, including the pyrochrome solution, are very unstable. Mix the two parts just before use, as after ten minutes or so, the developer is *dead* (useless). Naturally, these developers should be discarded when you're done.

With all procedures, the last step is to wash the finished hologram in running water and dry at room temperature. Be very careful concerning the wash, as the emulsions of holographic films are delicate and easily damaged. Also, the use of wetting agents, such as Kodak Photo-Flo, will greatly reduce the hazards of emulsion shifting and water spotting.

CHEMICAL	QUANTITY
DEVELOPER SOLUTION A	
PYROGALLOL	10 GRAMS
WATER	1 LITER
DEVELOPER SOLUTION B	
SODIUM CARBONATE	60 GRAMS
WATER	1 LITER
BLEACH SOLUTION C	
POTASSIUM DICHROMATE	4 GRAMS
SULFURIC ACID (CONCEN)	4 MILLILITERS
WATER	1 LITER

This chart gives the chemical make-up of the developer and bleach solutions for this process. The fixer solution is any normal photo fixer without a hardener. All or most of these supplies should be available at photographic supply shops.

The developer is a two-part arrangement, and the two solutions should not be mixed before you are ready to use them. The mixed developer will spoil in a matter of ten minutes or so, and should be discarded after use.

Addition of 1 to 25 grams per liter of sodium sulfite to solution A controls the color of the finished hologram from red to green.

Table 8-2. Pyrochrome formulas.

The Kodak approach is slower but much easier. It requires a three-step procedure of development, stop bath and fixing, much the same as with normal black-and-white film. As always, once processed, the hologram has to be washed and dried.

With any of these processes, the temperature of all chemicals and the water bath (wash) should be kept the same. This prevents such nasty things as emulsion cracking (reticulation) and emulsion shifts that result in image movement.

That's about all there is to it. It isn't hard to process your holograms, but one caution is in order. Like most silver-based films, image degradation will occur if the film is not processed promptly after exposure. However, with holograms this is critical. Even a day or two delay in processing will likely lead to vastly debased final product. So, plan on *running* your film as soon after exposure as possible!

On a final note, here. Large photo houses should carry holographic supplies, but with smaller outfits, you will probably have to order them. Thus, I have included a list of both Agfa and Kodak film and chemicals, labelled *Table 8-3*, for your convenience. This list is accurate as of this writing, but as stated before, changes occur rapidly.

HOW DOES ALL THIS WORK?

Well, it ain't magic!...although it sometimes appears as if it were. Actually, holograms are the result of applying some basic principles of physics. Basic, but not always simple in nature or simple to understand.

However, I will try to make this as easy to comprehend as I can. But, the fields of light and optics can be just plain confusing at times.

Anyway, remember that a hologram is the product of two beams of light striking and exposing the film. When these two beams get together, they set up an *interference pattern*—light interfering with other light, or itself. Since both these beams are coming from the same light source (the laser), they actually arrange the microscopic particles of silver in the film's emulsion as a series of *micro-mirrors*.

It is these *mirrors* that give the hologram its phenomenal property; that is, the object can be viewed from all angles, much in the same way that, as with any standard mirror, you are able to see beyond its physical boundaries. A hologram expands that principle to see completely around the object, and this 360-degree perspective is recorded on the film.

In way of review, this single light source is producing both the reflection and transmission beams. And, as the two strike the film's emulsion, each sets up its own interference pattern. These patterns then lead to the *mirror* effect that creates a hologram.

This is, of course, a very basic, and somewhat generic, explanation. There is a lot more to holography than just that, but if I delve much deeper into this, I'm likely to confuse you, and myself, for that matter.

AGFA-GEVAERT FILMS	
FILM TYPE	HOLOGRAM TYPE
10E75	RED LASER-TRANSMISSION
8E75HD	RED LASER-REFLECTION
AGFA-GEVAERT CHEMICALS	
CHEMICAL DESIGNATION	FUNCTION
GP62	REFLECTION DEVELOPER
GP61	TRANSMISSION DEVELOPER
GP432	BLEACH
GP334	FIXER

This list provides the films and processing chemicals available from Agfa-Gevaert. As can be seen, Agfa makes two different developers: one for transmission holograms and another for reflection holograms. Also, this list does not include the films Agfa makes for lasers that produce blue/green beams.

Table 8-3A. Hologram films and processing chemicals (Agfa-Gevaert).

Just remember that exposure methods used to make holograms place the entire image of the object on the entire plate or piece of film. For example, if you were to smash a hologram into many pieces, each individual piece would allow you to see the same complete image that was the original hologram. You will have to move the "piece" up and down and side to side for a full view, but it will be there.

CONCLUSION

Holography! Ah, what a world. After you have done a few of these, you will be "bit by the bug." And, there's no antidote! But, that's alright, as this can be a very rewarding hobby. Furthermore, it can be quite profitable. Many a local museum will pay top dollar for holographic images.

FILM TYPE	HOLOGRAM TYPE
120 PLATE/SO-173 FILM	RED LIGHT-TRANSMISSION
131 PLATE/SO-253 FILM	RED LIGHT-REFLECTION
649-F PLATES	RED LIGHT-REFLECTION
KODAK PROCESSING CHEMICALS	
CHEMICAL DESIGNATION	FUNCTION
D-19	DEVELOPER
INDICATOR STOP BATH	STOPS DEVELOPMENT
FIXING BATH F-5	FIXES FILM
HYPO CLEARING AGENT	CLEARS FIXING SOLUTION
PHOTO-FLO 200	PREVENTS WATER SPOTS

This list provides the films and chemicals available from Eastman Kodak. The D-19 developer will do all types of holograms, which does make things a little easier. Also, the developer has a shelf life of better than six months, instead of 10 minutes. It should be discarded after use, however. Like Agfa, Kodak also makes films specifically for blue/green light lasers.

Table 8-3B. Hologram films and processing chemicals (Eastman Kodak).

But, more to the point, this is a superb example of the versatility and merit of lasers. Few other devices can boast such a wide scope of applications. I mean, these things can do everything from cutting steel plate to making these pretty pictures. Well, almost everything.

To start making these 3-D images, you will have to exert a little effort, and spend a little *moollah*. But, the end result will be a fascinating and rewarding journey into a special realm of the laser world, one known as *holography*. And, I'll just bet you're gonna love it!

So what are you waiting for? Break out the tools and get to work on that positioning table. The sooner you finish it, the sooner you will be *turnin' and burnin'* on those holograms. Just think of all the people you will "WOW" with these little devils!

CHAPTER 9

LASER CONTROLLED SYSTEMS

In this chapter, we will talk about ways to use your laser(s) to control other devices—devices such as relays, silicon-controlled rectifiers (SCRs) and alarm systems. In many respects, this might be one of the most vital applications to come out of this book.

Through some modest circuits, the laser can be employed to turn stuff on and off or alert you to intruders in your home and/or on your property. Thus, it can become one of your best friends, providing you with safety and peace of mind.

Inherently, due to its collimated beam, the laser is a natural for these tasks. Since the beam maintains a nearly pinpoint spot over long distances, there is little problem in using mirrors to direct it, say, around corners. And, that aspect lends itself flawlessly to a number of projects.

So, I have divided this section into four areas: relay activation; SCR activation; internal security; and perimeter observation. Basically, these are all generic designs, but each has the potential for numerous applications regarding both convenience and protection.

Let's take a look at all four individually, and since there is nothing overly complicated about any of them, they should be easy to grasp. I have suggested certain duties for each, but I'm confident you will think of others.

LASER RELAY ACTIVATION

Employing a laser beam to trip a relay makes use of a relatively simple electronic technique; that is, utilizing U1, an operational amplifier (op-amp), as a comparator. Referring to *Figure 9-1* will acquaint you with this strategy.

As seen, the *inverting* op-amp input (pin 2) is connected between fixed resistor R1 and cadmium sulfide (CdS) photoresistor CDS1. This sets up a voltage divider that will change as the intensity of the light on CDS1 changes.

The reference is set by potentiometer R2, whose *trimmer*, or center lead, is connected to U1's noninverting input (pin 3). In this arrangement, when the laser is illuminating the photoresistor, the reference (threshold) can be set to keep the op-amp output low. However, if the light is removed, or blocked, the output goes high.

Said output is connected to the base of transistor Q1, through resistor R3, and when it goes high, Q1 acts as a switch letting the positive voltage pass from collector to emitter. That voltage then proceeds through the relay coil to ground, activating it—the relay (RLY1), that is. Diode D1 is used to prevent the relay's collapsing field from damaging Q1 and/or U1.

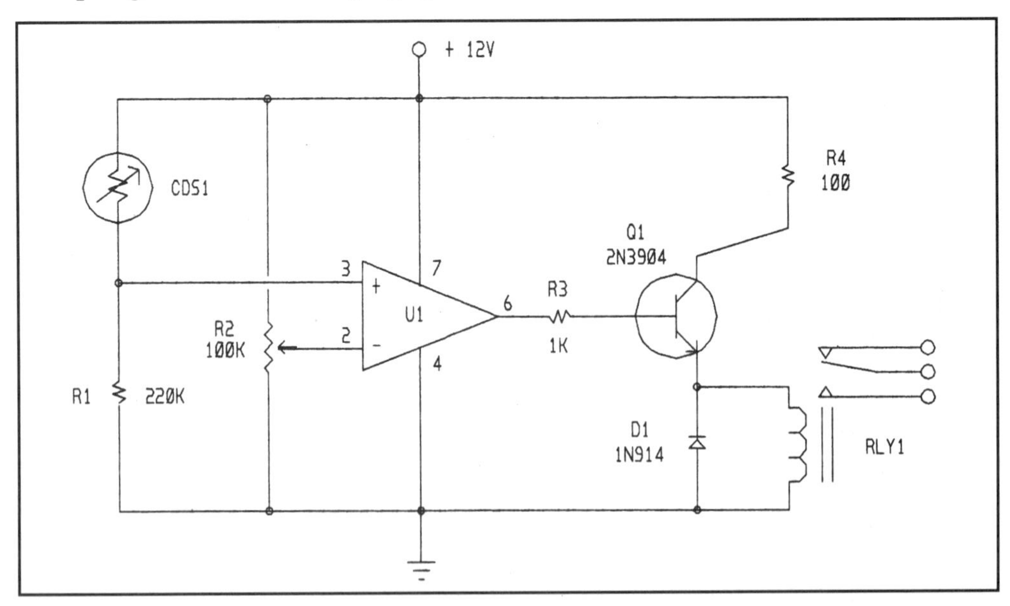

Figure 9-1. CDS sensor relay detector. With this circuit, the relay will activate each time the light beam is broken, and deactivate when the beam is restored.

When the illumination returns, Q1 is shut off, and the relay deactivates. So, the position of the relay contacts can be controlled by the presence, or lack thereof, of the laser beam. Naturally, this will also manage whatever device(s) you might have hooked up to those contacts.

By the way, if you want the system to respond to a sudden appearance of light instead of a sudden disappearance, reverse the IC inputs. Tie the inverting input to the CDS1/R1 connection and the noninverting input to potentiometer R2.

Since relays can be found with everything from SPST to 4PDT (sometimes even more) contact arrangements, your laser can do a whole lot of controlling if you want it to. It could turn a lamp on or off, activate an appliance, sound an alarm or all of those. That part is up to you!

As for construction, this can be done with either perf-board and point-to-point wiring or a printed-circuit board (PCB). I usually prefer a PCB, so I have included the foil pattern for this project as *Figure 9-2*. I strongly recommend an IC socket for U1, as, in case of trouble, this allows for IC replacement.

Install the discrete components first, then the relay, and finally the LM741. Incidentally, I have designated U1 as an LM741 op-amp, but virtually any single-operation amplifier will work. Just be sure that if you substitute the device, the one you choose has the same pinout as the 741.

With all that done, the unit is ready for service. Decide *what* you want to control, and connect that *what* to the relay contacts. The relay suggested for this project will handle about 5 amps at 120 volts. So, if you need to manage a device that requires more current than that, you will have to use this relay to trip a larger relay.

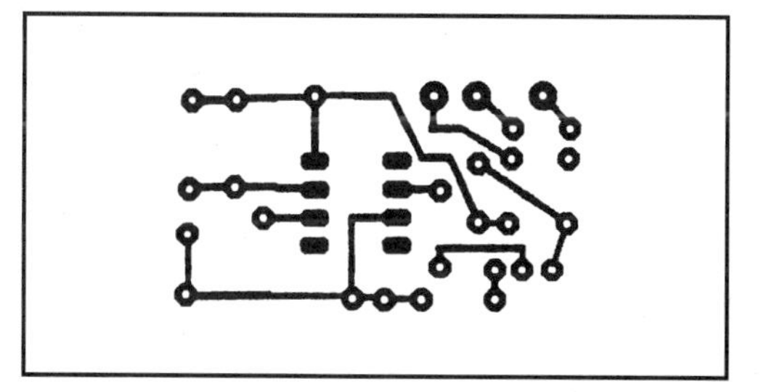

Figure 9-2. CDS sensor relay PCB pattern.

But, as they say, "It's all in a day's work!" Whoever "they" are. I never have been able to find out, and it's not from lack of trying, mind you!

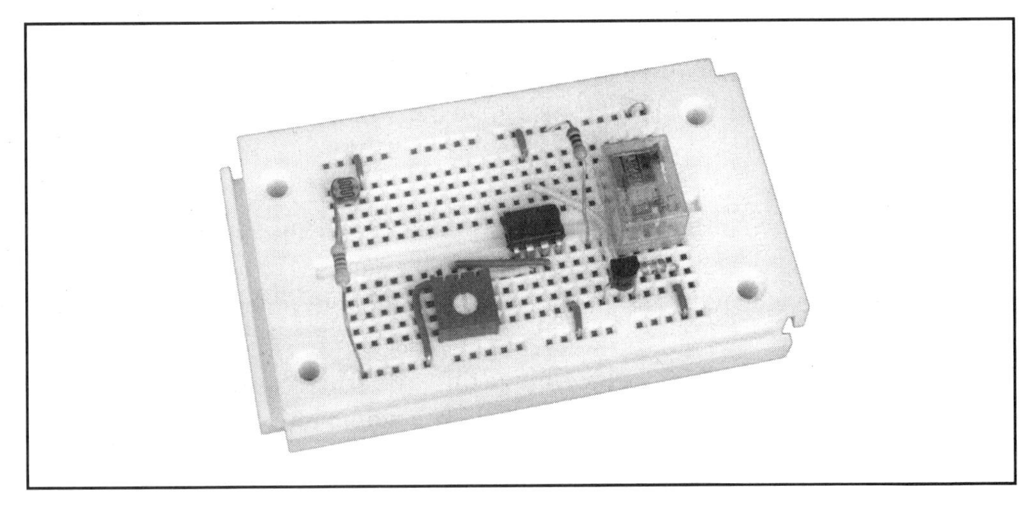

Photo 9-1. Breadboard version of the CdS detector relay control circuit. The potentiometer adjusts the sensitivity.

LASER ACTIVATED SILICON CONTROLLED RECTIFIER CIRCUITS

If you thought the last circuit was elementary, wait 'til you see this one. It is nothing more than a phototransistor, a resistor and the silicon controlled rectifier (SCR). The *load*, or device controlled, can be a relay or something else. With the *something else* being anything than doesn't need more voltage and current than the circuit can provide (a light-emitting diode (LED) for example).

Since this system is so fundamental, I recommend perf-board/ point-to-point technique (this is a real departure for me, as I'm a "dyed in the wool" PCB fan). This one just doesn't justify the trouble involved in making a printed-circuit board.

Figure 9-3 (shown with a relay load) illustrates just how simple the device is. The phototransistor (Q1) detects the laser beam and energizes SCR1 by applying a potential to its gate. Resistor R1 is used for biasing purposes, and switch S1 resets the circuit after activation.

They don't come much simpler than that. As previously stated, the relay could be replaced with a device that falls within the system capacity—in this case 12 volts. Or, by applying a separate power source, such things as direct current (DC) motors can be handled, provided the required motor voltage doesn't exceed the system potential, and a common ground exists.

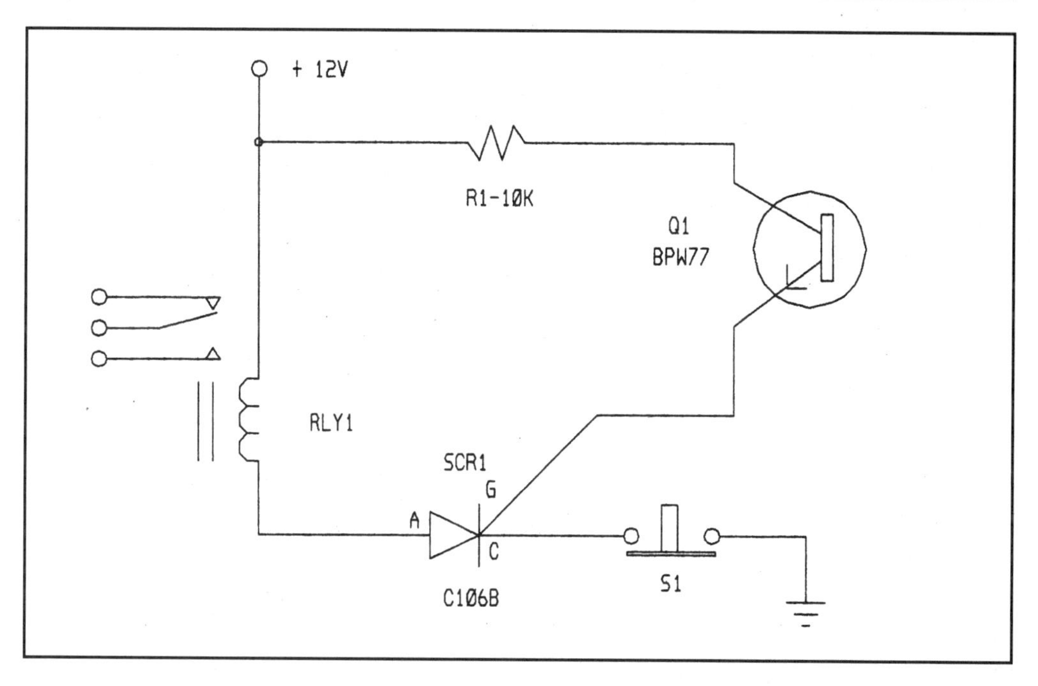

Figure 9-3. SCR based relay system. This circuit will activate and lock the relay every time a beam strikes Q1. S1 must be pressed to reset the system.

One drawback to this circuit is that it has to be reset by S1 after each activation. However, that also has its upside. Once energized, this system stays energized until power is interrupted. That, of course, makes it splendid for alarms.

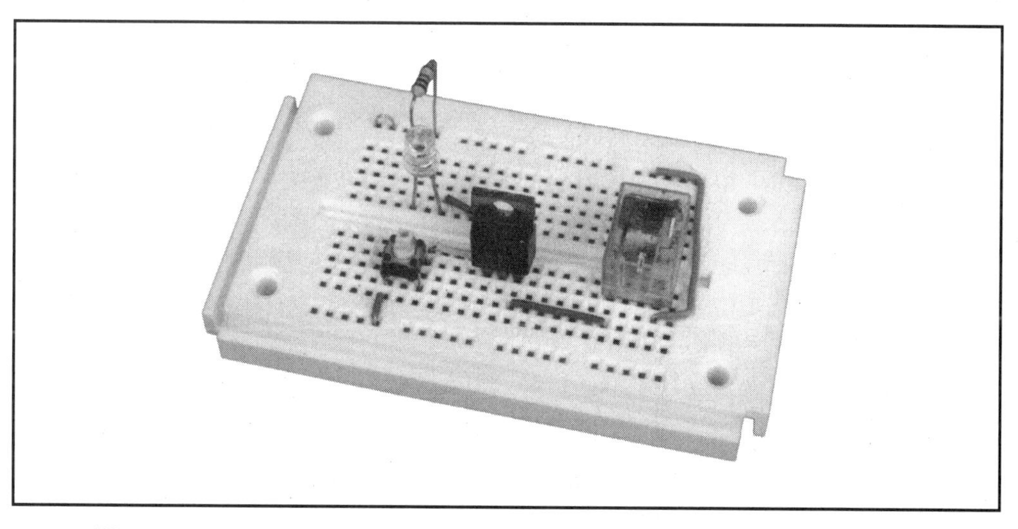

Photo 9-2. Breadboard version of the phototransistor/SCR latching relay control circuit. Note the simplicity of this system.

LASER CONTROLLED SYSTEMS PARTS LISTS

CDS SENSOR RELAY RECEIVER

SEMICONDUCTORS

U1	LM741 Operational Amplifier
Q1	2N3904 NPN Signal Transistor

RESISTORS

R1	220,000 Ohm 1/4 Watt Resistor
R2	100,000 Ohm Panel-Mount Potentiometer
R3	1,000 Ohm 1/4 Watt Resistor
R4	100 Ohm 1/4 Watt Resistor
CDS1	Cadmium Sulfide Photoresistor (Radio Shack)

OTHER COMPONENTS

RLY1	12-Volt DC PCB Relay

SILICON CONTROL RECTIFIER RELAY RECEIVER

SEMICONDUCTORS

SCR1	C106B1 Silicon Controlled Rectifier (SCR)
Q1	BPW77 NPN Phototransistor

RESISTOR

R1	10,000 Ohm 1/4 Watt Resistor

OTHER COMPONENTS

RLY1	12-Volt DC PCB Relay
S1	Normally Closed (NC) Pushbutton Switch

MISCELLANEOUS:

Solder, PCB Materials, Hardware, Cases (if desired), Collimator Materials (if used), IC Sockets, Hookup Wire, 12 VDC Power Supply, Etc.

OK, lets move on to a typical alarm system for your home or business. You will be happy to know, at least I think you will, that the two previous circuits are "key players" in this drama. And that, folks, makes things easier.

LASER BASED INTERIOR ALARM SYSTEM

Here is a method to protect a room inside a building using a helium-neon laser, some mirrors and one of the above mentioned detector circuits. And, it works right well. Actually, this system is very similar to the next project which involves guarding the boundaries of your property. So pay attention! Just kidding!

The principle here is to project the beam across the area you want to secure and have it end up at the detector. In this fashion, anyone (anything) moving in that area will break the beam and activate the alarm. Simple, eh?

Figure 9-4 demonstrates this principle...well, I hope it does, anyway. Of course, the configuration doesn't have to be exactly as the diagram shows. For example, the laser could be on the opposite wall from the sensor. The arrangement you select will probably depend largely on your circumstances.

One nice aspect of laser light is that unless you are looking almost directly into it, you can't see it—not unless the room is filled with smoke or some other reflectant. Hence, these systems are quite covert in nature. And, if you want to use infrared lasers, the system can be completely covert.

As a sidebar, though, infrared is a pain in the posterior to aim, especially when multiple mirrors are involved. So, bear this in mind if you're thinking infrared.

Now, back to our regular feature. Depending on your intentions, these alarms can be loud and noisy to scare the varmint off, or discrete so that whoever has triggered the alarm won't know that you know that they have done so. He, He, He!

Also, it is wise to keep the beam up off the floor a ways to prevent pets from tripping the system. How far up off the floor will depend to a great extent on how big your pets are.

Regarding the hardware, the mirrors can be either front- or rear-surface and should be as small as possible. Naturally, that's to keep them from being noticed. You

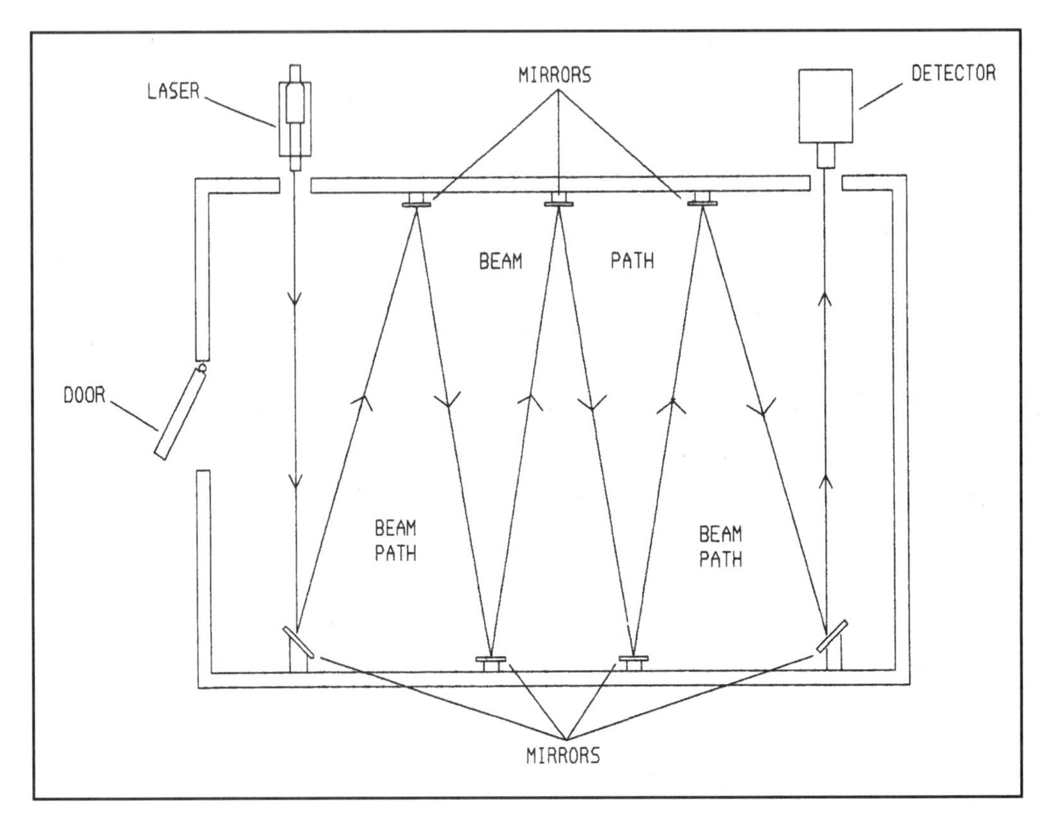

Figure 9-4. Interior laser based alarm system.

don't want them so small, however, that a vibration within the building, such as a slamming door, momentarily shakes them out of the beam's path. That would be interpreted as a blocked beam and trigger the alarm.

The rest of the system is up for grabs. There is spacious room for flexibility and/ or modification, which makes this project all the more desirable. And, with the right placement of mirrors, this system could also protect more than one room.

LASER BASED PERIMETER ALARM SYSTEM

Next! A perimeter alarm is just a variation of the last one. Only this time, the beam will leave the structure, travel around its boundary then re-enter the building to strike the sensor. Again, any break in the beam, such as some jerk climbing over your fence, will trip the bells and whistles.

Here, outdoor lights coming on and loud sirens blaring will probably be more effective than a discrete warning, but the alarm can be done covertly if desired. Either way, you will know someone, or something, has breached your property line.

In this scenario, mirrors are used to first direct the beam outside the building, usually through a window, around the path of the perimeter then back in through the window to the detector. One problem that plagues this type of system is the effect of the elements on the mirrors.

They have to be sheltered as well as possible from rain, sleet, snow, and so forth, and even then will have to be replaced from time to time. The silver coating on even the best mirrors is no match for what the weather has to offer. And unfortunately, this is true no matter how well you protect them.

Even so, it is well worth the effort, considering all the security and peace of mind this system will bring you. I mean, what is the welfare of you and your family worth? Besides, the mirrors are cheap enough.

As for the proper method of protection, little tiny plastic house-like structures, with an appropriately placed hole to let the beam in then back out, usually do the trick. By placing slanted pieces (roofs) on top to drain off water, the mirrors are rarely subjected to a destructive amount of moisture.

Of course, if the rain is coming in at a very steep angle, there is little you can do about that. And humidity is another story. Again, not much can be done to protect against it.

When mounting the mirrors in their little houses, it is best to do so with flexible wire. In this fashion, the mirrors can be easily adjusted to the required angles. And, the unavoidable periodic readjustment is not as big a problem.

Once you have all the mirrors in place, it is time to line everything up. Turn on the laser and position the first mirror so that the beam reflects off it and strikes the second mirror. Next, arrange the second mirror so that it strikes the third. Now, move around the perimeter adjusting each successive reflector as you go along.

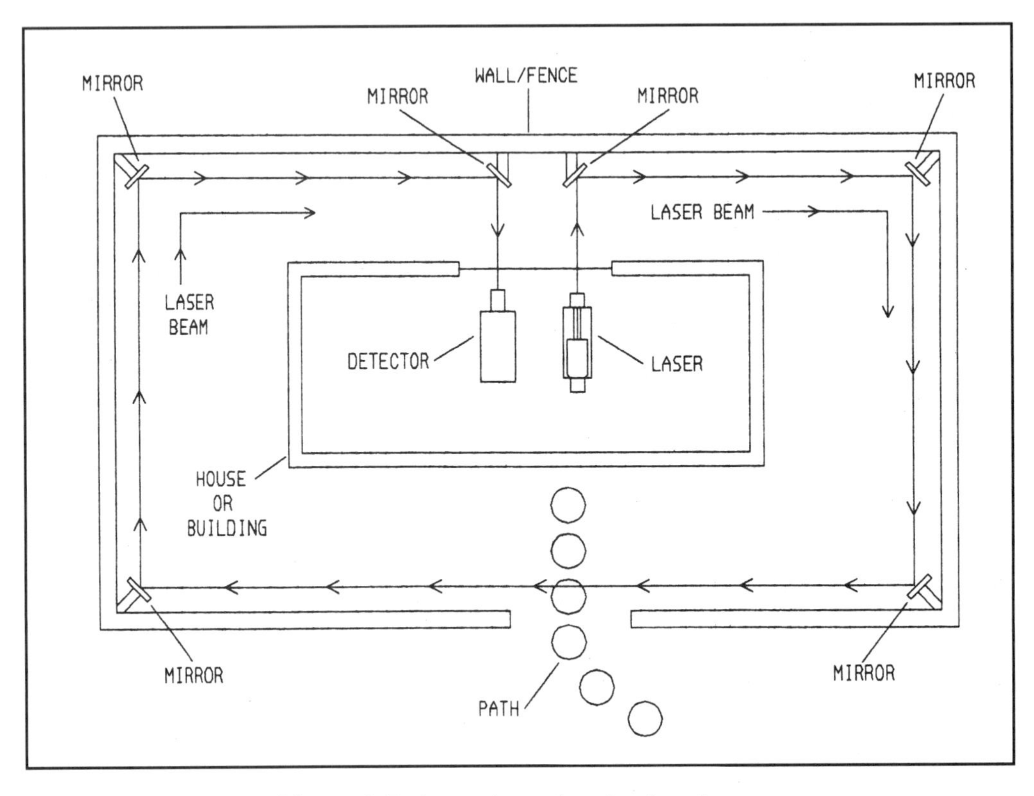

Figure 9-5. Laser based perimeter alarm.

Eventually, you will end up back at the window. The last step is to shine the beam in and on the detector. If all is well, anybody/anything that tries to come past the boundary line will break the beam and sound the alarm.

And that, as they say, is that. (There's "they" again.) Anyway, following this procedure will result in a well protected perimeter. As mentioned before, without something to reflect off of, a laser beam is next to impossible to see. Thus, intruders are rarely able to beat your system because of a visual warning of its presence.

CONCLUSION

Well, gang. There you have the laser controlled relays and alarms. I hope this information will be useful, and perhaps spawn some ideas of your own. Personally, I never cease to be amazed by what can be done with lasers.

But, some of the very best applications are those that provide safety and protection. And, these projects certainly fall into that category. To that extent, they are well worth the time, effort and expense it takes to assemble them.

I have always enjoyed exploring the properties of light, as well as what could be done with it. This might explain my early interest in lasers. Well, then again it might not. But one thing is for certain. These projects definitely demonstrate some of the more fascinating attributes of what we all know of as *light*.

Be it visible or invisible, the magnetic spectrum that makes up this phenomena has a personality all its own. It refuses to bend but can be reflected endlessly. It can span billions of miles of outer space, yet can be stopped in its tracks by a particle of dust. And, like an electron, you can't touch it, smell it or taste it, but its presence is glaringly clear. Pardon me!

But, the best part of all this is how we, as mere human beings, are able to use light to such a great advantage. While all illumination is multifaceted, lasers embody light in possibly its most versatile form. So, build these projects and put this marvel to work for you. And don't forget to have some fun along the way!

CHAPTER 10

A SEMICONDUCTOR LASER DIODE SYSTEM

Now that we have experimented with the gas based helium-neon laser tubes, and have those mastered (We do, don't we?), let's play with one of the newer additions to the laser world—the semiconductor, or *diode* laser. As you know by now, the HeNe tube is also a diode, but our next endeavor will deal with a specialized version of the light-emitting diode (LED) that produces true laser light.

As with everything electronic, there are some compromises associated with the diode laser; that is, some good things and some not so good things. However, the good things definitely outweigh the not-so-good, at least concerning certain applications.

Probably the most significant "good thing" is the system power requirement. Unlike the dangerous high voltages demanded by gas lasers, a diode laser runs off potentials in the 2.5 to 6 volt range. This, I think, speaks for itself.

A second advantage involves size. These little jewels come in two standard packages, with diameters of 5.6 and 9 millimeters. Very small, indeed. And, if surface-mount technology (SMT) is employed, the transistorized driver circuits can be as small as 10 millimeters square. So, it is not hard to see how these laser systems easily fit in a case the size of a fountain pen (a.k.a. the popular "laser pointers").

A drawback is that while the voltage is low, the diodes tend to use about 10 times the current of their gas tube cousins. This equates to around 30 milliamps (mA) for threshold and 35 to 40 mA for the operating current. But, even at that, battery powered applications are both common and efficient.

Another "not so good" aspect of the laser diode concerns *collimation*. Due to their size and architecture, these devices do not produce a beam that even approaches the tight *spot on the wall* profile encountered with other lasers. This "flaw" is easily corrected with optics, but it does require the addition of a collimating lens assembly.

So with all that to think about, let's try our hand at a diode laser system. This project will acquaint you with semiconductor lasers and result in a working visible light unit that should come in handy regarding a number of future ventures.

THEORY

Since I covered the structure and functionality of a laser diode in Chapter 1, I ain't a 'gonna get redundant and cover that again. Instead, we will look at the driver circuit and how it allows the diode to perform as a full fledged laser. By the way, *Figure 10-1* illustrates the two most common internal configurations for laser diode modules and the most common pin assignment.

Naturally, there are dozens of different designs available for driver circuits. Some use transistors, while others employ operational amplifiers (op-amps), but in the past, I have found that transistors usually do as good a job, if not better, than op-amps. So, our system will use two transistors.

Normally, diode laser modules contain a photodiode to monitor the laser diode's output. This is done to keep the light consistent, but more importantly, it protects the sensitive internal gallium-arsenide junction from excess current damage. These junctions can burn up very rapidly from too much current.

Referring to *Figure 10-2*, the schematic diagram, note that transistor Q1's base is connected to the photodiode anode pin of the laser module. In this fashion, the photodiode controls the output of Q1 which is connected to the base of Q2.

Since Q2 governs current flow to the laser diode, the photodiode indirectly supervises this as well. That is, as the current to the laser increases, its light intensity also increases. That, in turn, causes the photodiode to decrease Q1's base current which readjusts Q2's base current and the laser diode output.

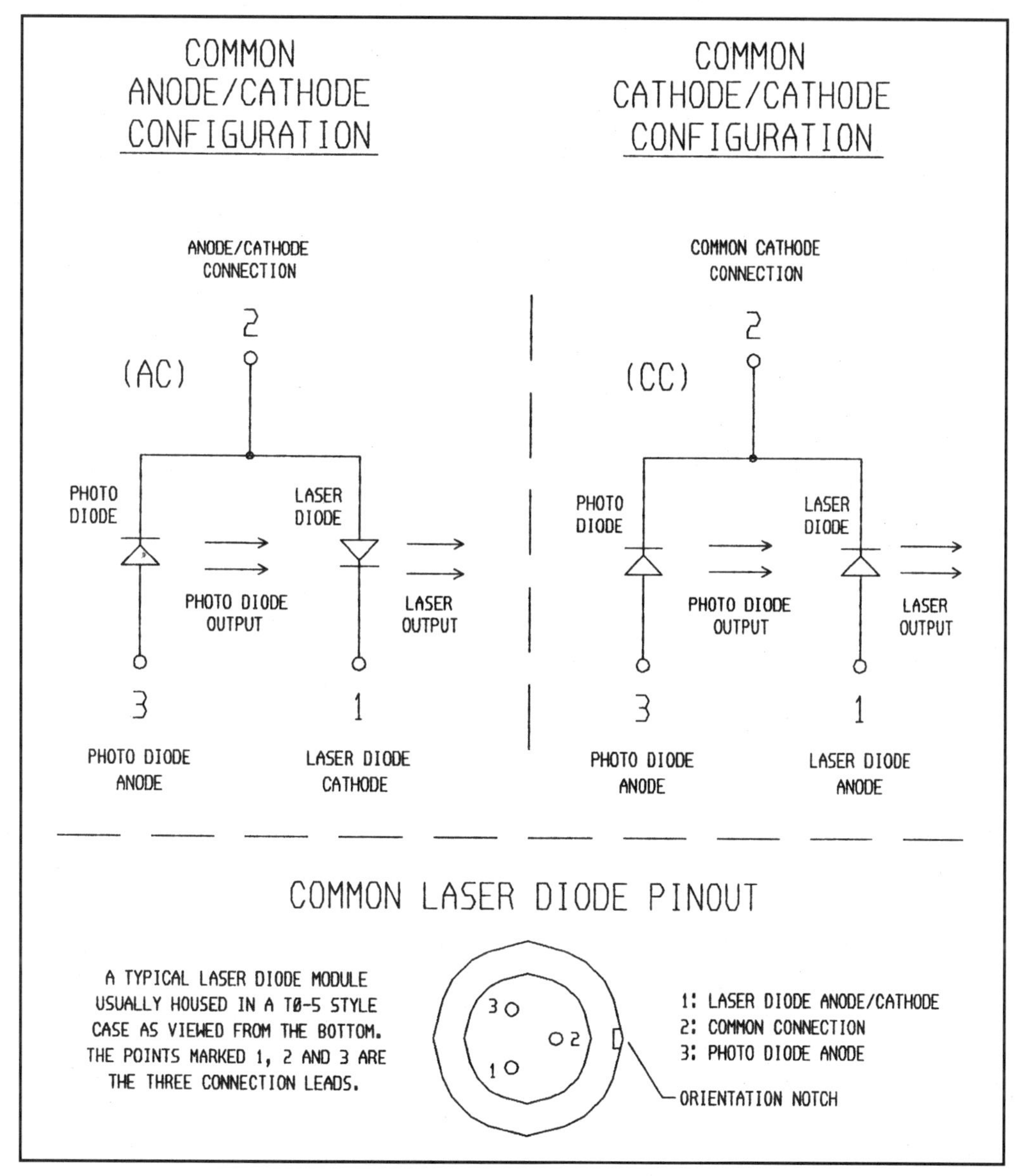

Figure 10-1. Laser diode layouts.

You got all that? Good! Now, explain it to me! Just kidding, I'm not completely lost yet. I know that gets a little confusing, but I can't help it. That's just the way the darn things are made.

OK, OK! Let me try this. The photodiode is included to monitor the light output of the laser diode. If the light gets too high, the photodiode turns the current down

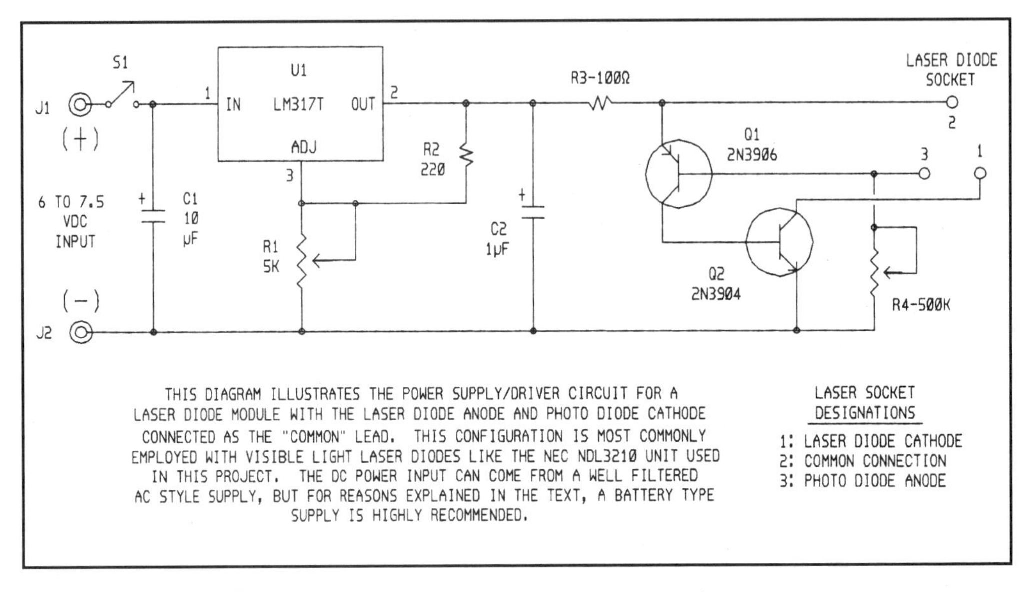

Figure 10-2. AC style laser diode driver.

so the laser doesn't burn itself up. Maybe I should have said that in the first place, but then you wouldn't know how the driver worked.

Anyway, as can be seen, the driver section is quite simple, consisting of only two transistors, a capacitor, one resistor and a potentiometer. The potentiometer is used to tailor the system to a variety of different laser diodes, but I will get to that in the testing section.

This driver is designed to handle lasers that have the photodiode cathode and laser diode anode connected together as the *common* pin (designated AC). For visible light diodes, this is the customary arrangement. Now, that doesn't mean they are all constructed that way. *Figure 10-3* shows a *cathode-cathode* (CC) hookup that is more often found with infrared lasers. However, you may well encounter either configuration with either wavelength. Be sure to confirm which you are dealing with from the data sheet.

Regarding the power supply, it is a standard linear arrangement using the LM317 (U1) adjustable positive voltage regulator. Resistor R2 and potentiometer R1 form a voltage divider network that controls the output voltage.

As with all regulators, the 317 requires an input voltage higher than its intended output. So, input DC is rated at between 6 and 7.5 volts. This will provide an

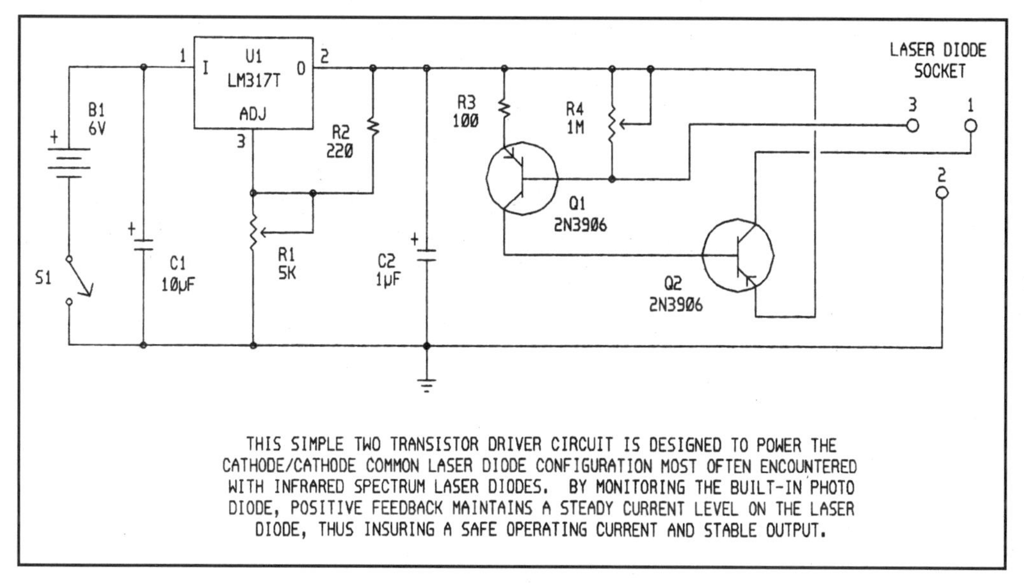

Figure 10-3. CC style laser diode driver.

input voltage of between 4.5 and 6.5 volts to the driver section, which should cover just about any visible light laser diode you encounter.

Connect the sections together (not literally, that's done for you by the PCB), and you now have a system that will light up those little varmints. Add some optics, and a small red dot, nearly as nice as the one your helium-neon laser throws, can be projected on whatever "target" you choose. Don't panic! I'll get to the optics part in the construction section.

On a final note concerning power, these drivers will function from either AC based supplies or batteries. However, there is a liability to using AC. The module junctions are not only susceptible to excessive current but to transient line signals and/or voltage spikes as well.

Considering that both appear regularly on standard AC household lines, there is a definite risk associated with powering your laser from that source. Hence, I highly recommend battery packs. With those, you will not have to worry about either "gremlin" frying your precious, and expensive (well, not so expensive these days), laser diode module.

Alright, there it is. In review, the linear supply provides a regulated voltage, and the drive circuit employs feedback from a photodiode to maintain a steady and

Photo 10-1. Closeup shot of the laser diode driver circuit. This two-transistor unit has adjustable current gain.

safe current level. Hey, what more can you ask for? Never mind! Don't answer that! Let us move on to constructing this marvel of modern electronics.

CONSTRUCTION

A quick word of caution before we move on. Diode lasers are also very sensitive to static charges and must be handled with appropriate caution. Such things as grounded wrist straps and a grounded work area go a long way towards protecting these devices when you are handling them.

Also, when storing the modules, do so in a container that is lined with anti-static foam, such as seen in *Photo 10-2*. This not only shields against possible static damage but helps keep the delicate lasers from rattlin' around inside the container.

OK! Back to business! Referring to *Photo 10-3*, you will see the completed assembly mounted on a clear 1/4 inch thick plexiglass base measuring 3 7/8 inches (98 mm) by 2 1/2 inches (63 mm). Of course, this system could be considerably smaller, but I arranged it as such to provide better comprehension of the overall system. Also, if someone wants to use it for a science class/fair project, this configuration tends to be more impressive. See, I watch out for you guys!

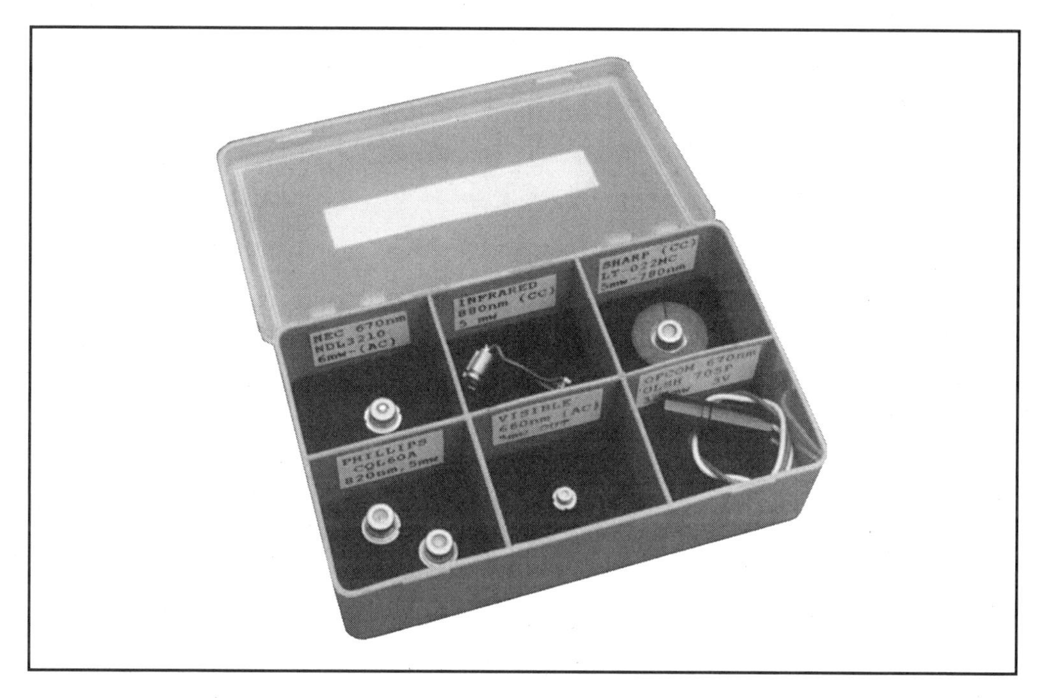

Photo 10-2. A simple and safe way to store laser diodes. Note the black conductive foam floor. This protects the diodes against both static charge and vibration.

Anyway, the first step is to build the driver. It can be done with perf-board and point-to-point wiring, but I'm a printed-circuit board (PCB) fan (freak?) and recommend that approach. To that end, I have included *Figures 10-4 and 10-5*, the foil pattern and parts placement diagram, to facilitate PCB construction.

With the finished board in hand—and it is easy to make—start with the small components (resistors/capacitors). Next come the potentiometers, regulator and transistors. I left some copper on the board around the regulator to act as a heat sink.

Now, the laser diode socket. This consists of a small piece of PCB blank with a triangular arrangement of three round pads (I positioned the triangle to point *north*). To each pad is soldered a single in-line pin (SIP) socket and an appropriate length of hookup wire.

This PCB is now cemented to a small piece of 1/4 inch thick plexiglass (the *stage*), with a 3/8 inch hole drilled for the leads, which is cemented to the end of the main board. I used "superglue" for both, but epoxy or PVC plastic cement would also work just fine.

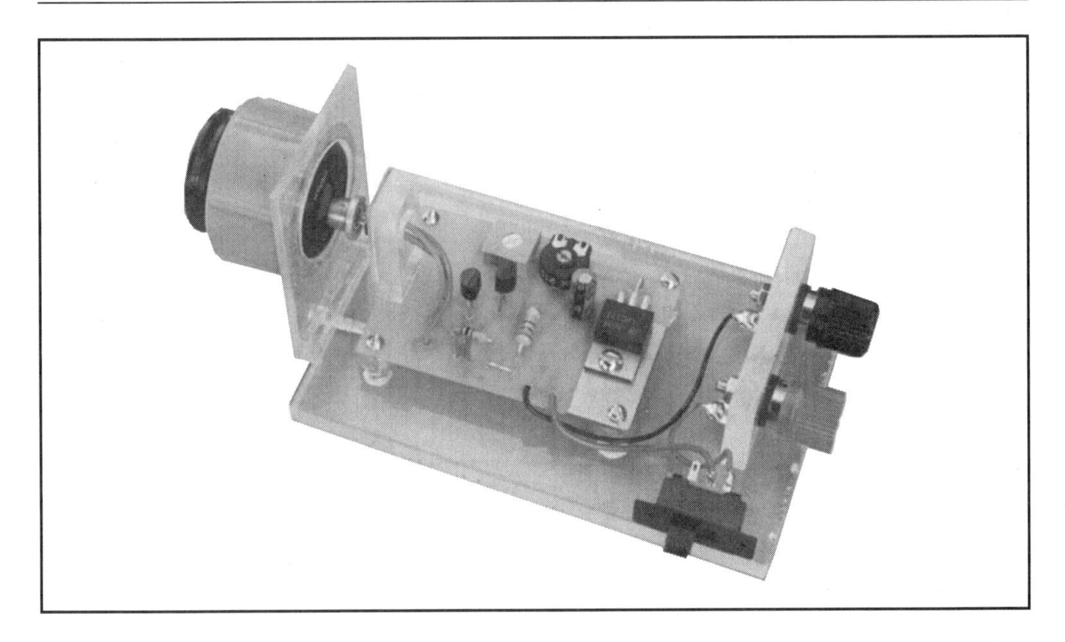

Photo 10-3. The completed diode laser system with collimating lens. Note: this unit could be smaller, if desired, but this size does help the user visualize the circuit.

With the socket/stage in place, the leads are soldered to the main board pads. Double check to be sure you have the right leads going to the right pads, as a mistake could ZAP your laser module.

Next, using some more 1/4 inch thick plexiglass, I fashioned a vertical piece to support two banana jacks (DC power). The negative jack is connected directly to the main PCB, while the positive post goes to switch S1, then to the PCB. That, naturally, is the "ON-OFF" switch.

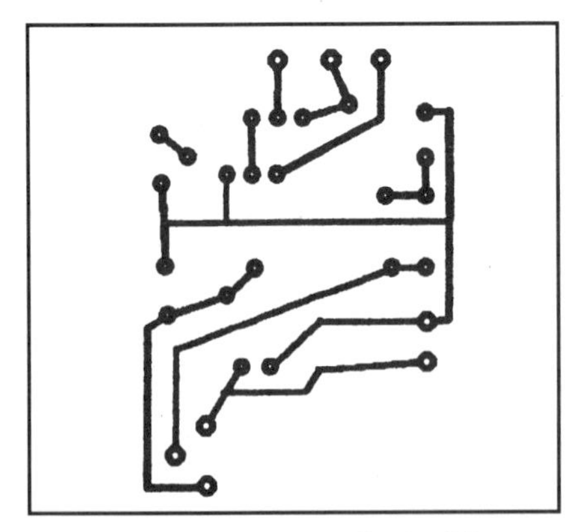

Figure 10-4. Laser diode driver foil pattern.

Again, superglue was employed to secure both the plexiglass and the switch to the base plate. The switch is of the *slide* variety, laid on its side—and be careful with the glue! Use it sparingly, as you don't want that stuff to get inside and glue the switch in one position. I did that once, and it didn't make me happy! I won't go into detail regarding what it did make me, as this is a family book company and it wouldn't be appropriate.

Alright, enough about that! With all connections made to the main board, it can now be secured to the base plate with bolts and spacers. That oughta keep it in place.

Now that those steps are completed, it is time to turn our attention to the optics. Uh, yeah, "da optics." Actually, there are a number of sources for a suitable lens system. A simple *magnifying glass* (double convex lens) will work, but these usually result in a flaring effect around the projected spot. Hence, you really need a lens system made up of more than one element.

These can be found all around you. For example, eyepieces for telescopes and microscopes, monoculars, camera lenses and projector lenses all fit the category and are applicable for this purpose. For the prototype, I used a surplus job that began life as the objective lens of a microfilm viewer. It has a short focal length, is *fast* and was CHEAP (about a dollar). Yep, it fills the bill right nicely.

So, lenses aren't as hard to find as you might think. In an earlier diode laser project, I borrowed (stole) the lens out of one of those "disposable" cameras you get at the drug, discount or grocery store—you know, the kind where you shoot the pictures, then send the whole works to the film processing lab.

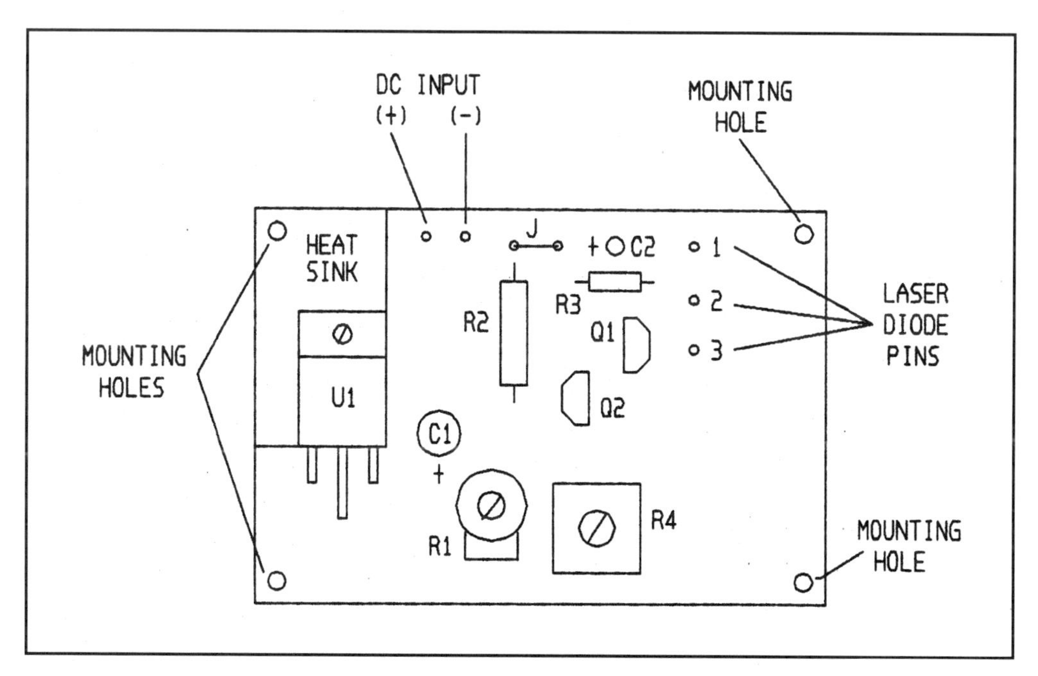

Figure 10-5. Laser diode driver parts placement.

This worked quite well, as the lens was mounted in a threaded bracket that allowed for focusing. However, it was designed for 35 mm film, so the focal length was about an inch. And that created a trivial problem where the edges of the beam fell outside the back lens element. This resulted in a slight loss in output, but so slight that it was hardly noticeable. You might want to keep this one in mind!

And, don't forget the surplus market! That is where I found the microfilm lens, and as was mentioned in Chapter 4, it is full of all sorts of optics, not to mention other "stuff" handy for laser hobbying.

OK, with my trusty little lens in hand, I needed to find a way to mount it—a way that would allow it to be focused. The lens has threads but doesn't come with the matching threaded mounting. So, I journeyed to my local discount hardware, lumber, and mostly assorted *junk* store.

There, I looked through the hardware section for a nut large enough, and with the proper thread, to fit my lens. Unfortunately, the only *nut* I found was a clerk who didn't seem to know where anything was. Not to be deterred, I put the ol' *gray*

Photo 10-4. Closeup view of the laser diode in the driver's socket. Nothe the proximity of the diode to the collimating lens.

matter in gear and set out to discover another source of threads. No, no...not clothes. Threads to fit the lens.

Next stop was the electrical section, and there....*and there*, I found a fitting for some sort of electric junction box that—you guessed it—was just right. As a bonus, it was made of plastic which also made it easy to cut and glue to some more plastic. Heck, this was my lucky day. I didn't even have to do battle with the store's "junk" department.

I promptly purchased a couple of the fittings and gleefully skipped home to live happily ever after. The End! That's all folks! Well, actually this elaborate, and

LASER DIODE SYSTEM PARTS LIST

AC STYLE DRIVER CIRCUIT

SEMICONDUCTORS

U1	LM317T Adjustable Positive Voltage Regulator
Q1	2N3906 PNP Transistor
Q2	2N3904 NPN Transistor

RESISTORS

R1	5,000 Ohm PCB Potentiometer
R2	220 Ohm 1/2 Watt Resistor
R3	100 Ohm 1/4 Watt Resistor
R4	500,000 Ohm PCB Potentiometer

CAPACITORS

C1	10 Microfarad Electrolytic Capacitor
C2	1 Microfarad Tantalum Capacitor

OTHER COMPONENTS

J1, 2	Banana Jacks
LDS	3 Single In-Line Pins (SIPs)
S1	SPST Slide/Toggle/Etc., Switch

(Continued next page)

LASER DIODE SYSTEM PARTS LIST (CONTINUED)

CC STYLE DRIVER CIRCUIT

SEMICONDUCTORS

U1	LM317T Adjustable Positive Voltage Regulator
Q1, 2	2N3906 PNP Transistors

RESISTORS

R1	5,000 Ohm PCB Potentiometer
R2	220 Ohm 1/2 Watt Resistor
R3	100 Ohm 1/4 Watt Resistor
R4	1,000,000 Ohm PCB Potentiometer

CAPACITORS

C1	10 Microfarad Electrolytic Capacitor
C2	1 Microfarad Tantalum Capacitor

OTHER COMPONENTS

S1	SPST Slide/Toggle/Etc. Switch

somewhat boring, tale of my quest to find the perfect lens mounting does have a purpose. And, it's not to fill up space, either.

It is meant to illustrate a method of conquering one of the foibles of prototyping: how to find parts. It has taken me years, but I have finally unlocked some of the secrets of this *fine art*. One is that often you will locate those parts in some of the most unlikely places. Auto parts stores, arts and craft places, building supplies and the like all offer possibilities for those seemingly impossible-to-find components.

A second trick isn't really a trick. It just good ol' fashion perseverance. If one prospective source bombs, keep trying. Nearly everything you need is out there, somewhere. Example, the plastic beads sold in arts and craft store to make necklaces also make great standoffs for circuit boards.

OK, let's move on. Now that I had my mounting hardware, I whacked it off to about 5/16 inch (with my trusty hack saw) to accommodate the lens. Next, I drilled a 7/8 inch hole, for lens clearance, in a thin piece of plexiglass and superglued the fitting over the hole. This assembly was then lined up with the diode socket and anchored to the front edge of the base with small screws.

And, that completed construction of the laser diode unit. If you have come this far, you too are done. And, it's time to plug in a laser diode and test the system. Well, almost time.

TESTING THE DRIVER

Before we place the laser, let's run a couple of quick power tests. Apply a 7.5-volt battery pack to the banana jacks and check the output at ground and capacitor C2's positive side. A reading of between about 1.2 and 6 volts should be present. Now, adjust potentiometer R1, and the voltage measurement should change. If it does, the power section is functioning correctly.

Next, test the current between pins 1 and 2 of the diode socket. These are the laser diode connections, and readings from almost zero to around 50 milliamps should be observed as you adjust potentiometer R4. If that checks out, the driver is providing acceptable current to operate most semiconductor lasers.

Now you can put the laser diode in its socket. Be sure the right pin goes in the right socket. With the prototype, as earlier stated, I arranged the #2 pin as the top point of the triangle, but that is merely by choice. The diode will work in any position as long as the connections are correct.

With that done, set both potentiometers to their center positions and flip the power switch (S1) on. You will probably see some glow from the diode. As with the HeNe tubes, don't look directly into the case opening. View it at an angle. Now adjust R1 for a voltage output of around 3 volts. Next, adjust R4 for the brightest laser light output.

You may want to play a little with both settings to achieve optimum performance, but the previously described procedure should get you very close. And, that is about all that needs to be done with the driver section of this system.

OPERATION

Now that you have your laser working, all that is left to do is to aim the little "beast" and focus the beam on whatever surface it hits. That, of course, is merely a matter of rotating the lens until the beam closes down to a bright red spot.

Note the speckling effect of the beam. That is a sure indication of true laser light. Incidentally, in theory, this beam can be focused to a diameter of its own wavelength. That is, if you are using a 670 nanometer diode, it should be possible to focus the beam to a diameter of 670 nanometers.

Naturally, that IS theory and to my knowledge has not been demonstrated in actual laboratory and/or field settings. However, they have come close, and it's that accomplishment that makes CD-ROM drives possible.

The data is placed on CDs by "burning" small holes (pits) in a sheet of metal foil. With 750 million bits of information on a standard CD, those pits have to be mighty small. And, only a light source capable of focusing to such a fine point (spot) is suitable for reading those minuscule individual data bits.

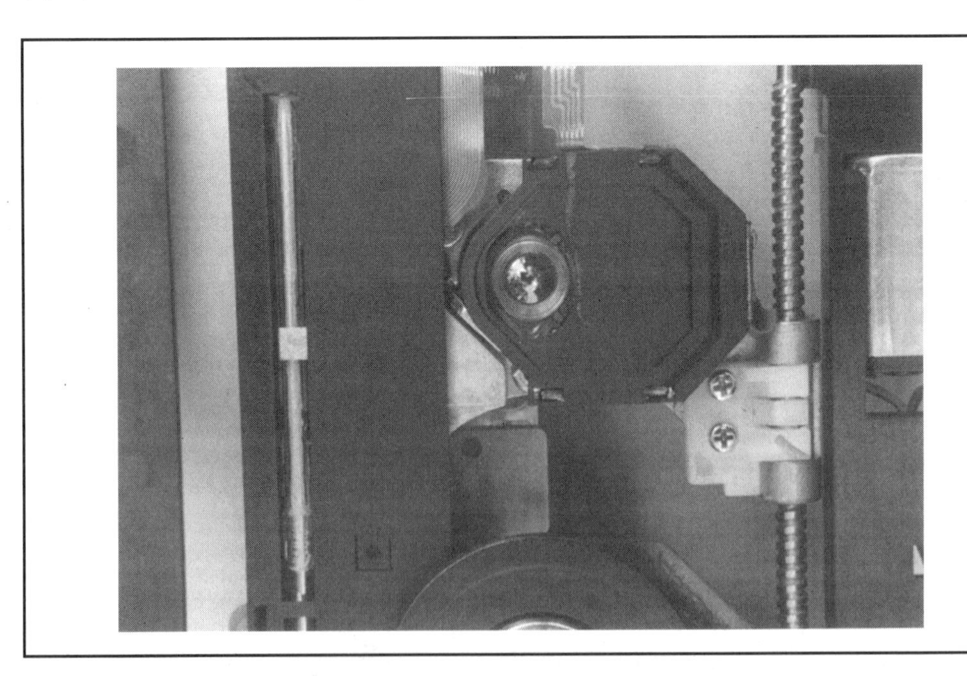

Photo 10-5. Closeup view of the laser/lens assembly of a typical CD/CD-ROM drive. Note the small collimating lens at the top.

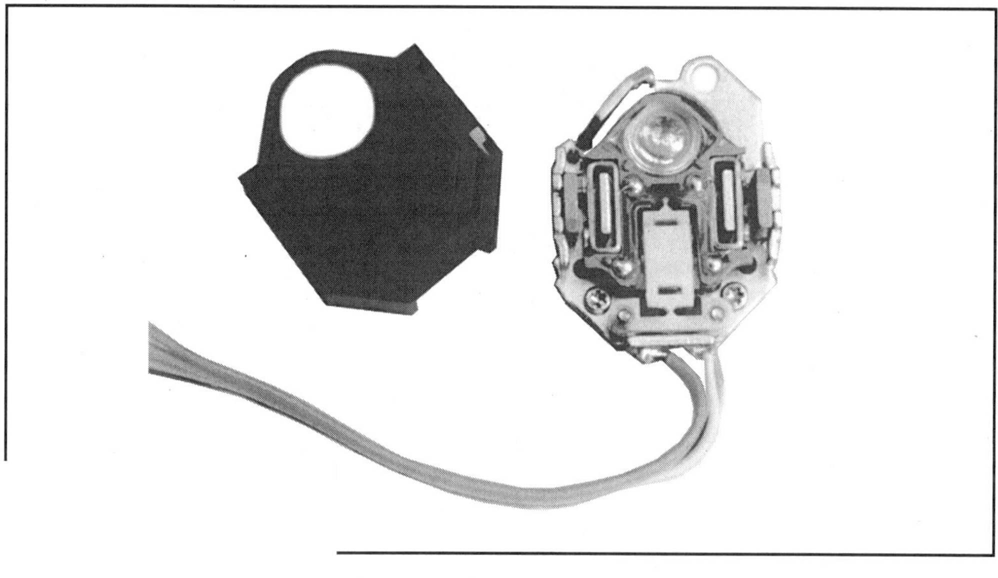

of the laser/lens assembly. Note the small coils on each
ˉhese are used to focus and position the lens.

ow ready to be put to task. Any of the experiments we
ı-neon tubes can be duplicated with this system. Just
; have to be focused to produce that small diameter
d with the HeNe devices.

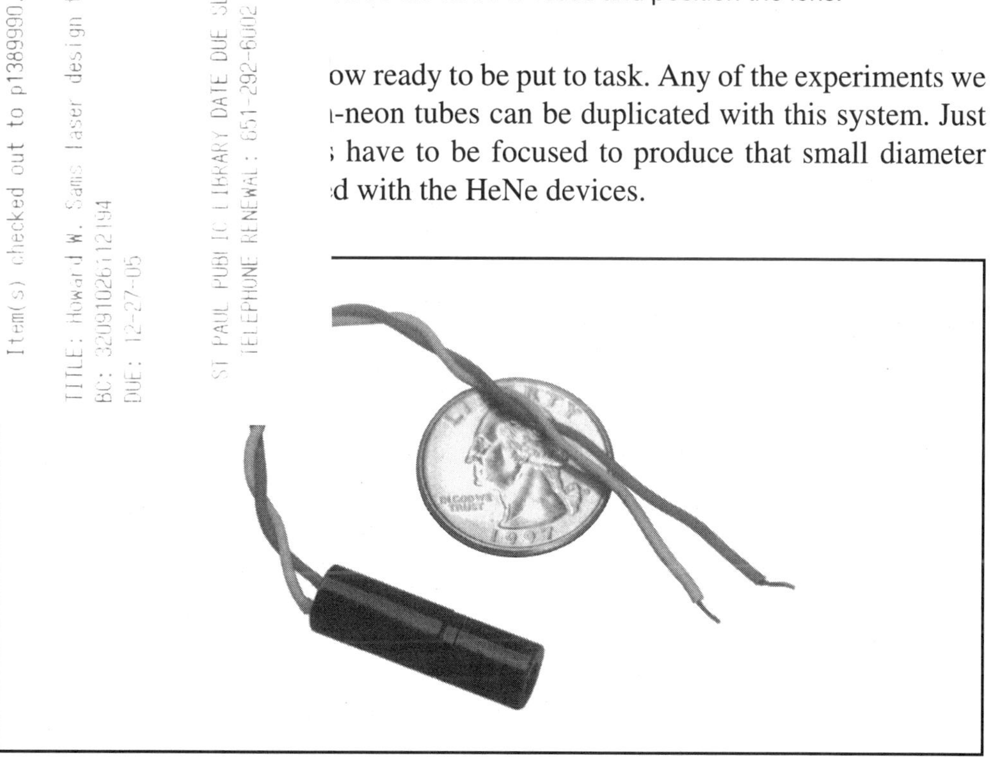

Photo 10-7. A typical laser diode module. These consist of the laser diode, driver and collimating lens. Note its small size compared to the quarter.

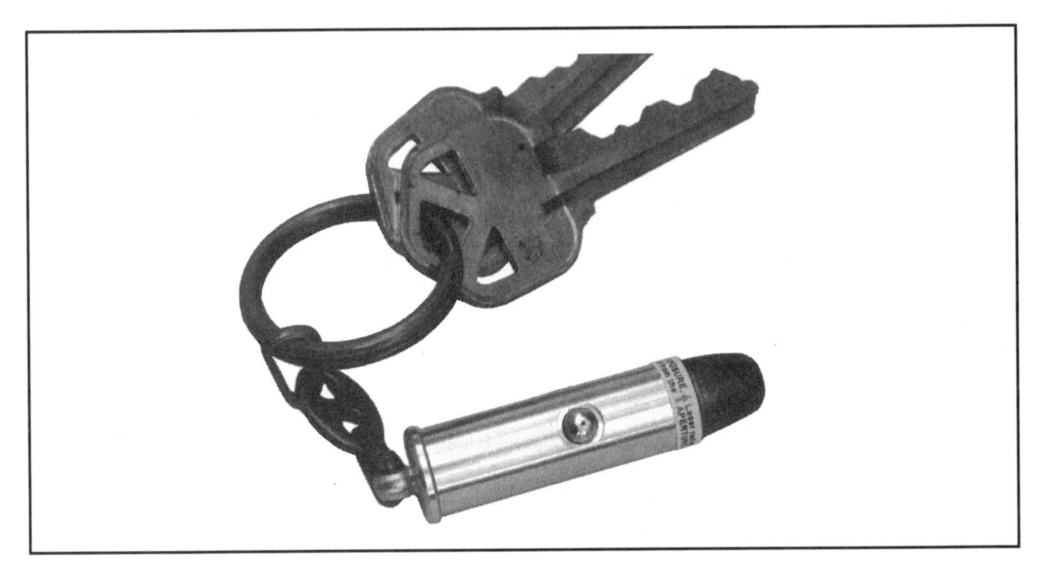

Photo 10-8. One of the newest laser pointers. Shaped like a bullet, this unit has a 650 nM diode and makes a handy key chain.

CONCLUSION

Over the years that I have worked with lasers, I'm always amazed at how this field keeps expanding. Perhaps I shouldn't be, as progress does march on. But, regarding this arena, someone is perpetually conceiving a new approach or device that is, at least in my mind, nothing short of astounding.

And so it was with the introduction of laser diodes. The initial infrared units didn't "yank my chain" all that much, but when the visible light devices arrived, WOW! Those were something else. Imagine, laser light coming from a package the size of a TO-18 transistor.

For that matter, I'm still fascinated with these semiconductor marvels. I can't help but view them as technology at its best, and with new wavelength and power ratings emerging practically every day, this aspect of lasers promises even more enchantment.

On that note, let me wrap this up. But, I do hope this chapter, and project, instills in you the same appreciation I have for laser diodes. When you consider the size and power requirements of these gems, possibilities for application are virtually unlimited. Put your imagination to work and have some fun with this one!

CHAPTER 11

ADVANCED LASER SYSTEMS

Buckle up folks! This is a long one, but I hope it will be informative. In this text we have dealt primarily with helium-neon and semiconductor lasers. However, there is a whole world of lasing devices out there you definitely need to be aware of. From time to time, I have mentioned some of these, but due largely to complexity, have not gone into great detail. So, I wish to correct that oversight in this chapter.

As stated earlier, theoretically, if enough of the right kind of energy is applied, virtually any substance can be made to lase. You may well destroy said substance in the process, but it will emit photons in the visible and/or invisible spectrum. And, it is this law of laser physics that has led to the development of such a vast number of different devices.

Now, I'm not suggesting you get a wild hair and set out to build these other systems. They would be both arduous and expensive to construct, with the end result often being quite dangerous. However, to fully appreciate the cosmos of lasers, it is advisable to have some knowledge of their existence. Besides, many of them are downright interesting!

Standing on that introduction, let me acquaint you with a selected number of the *players* in this game. Once you meet them, I think you will better understand the importance lasers play in our lives, and the versatility they exhibit.

OTHER GAS LASERS

Since lasers are classified in four major groups (Gas, Liquid, Solid-State and Semiconductor) and gas lasers account for the largest group, let's look at this realm first. Being aware of the physical properties of gas (plasma), scientists considered this medium the most likely for lasing action.

As was seen with the helium-neon device, current is passed through the gas mixture and results in what is commonly called *discharge excitation*. But this is not the only way to excite a gas. You can tell it a funny joke! No,...just a little humor (very little!).

Gas will respond to energy from a variety of sources, such as a chemical reaction within a gas (chemical excitation), exposure to radio frequency (RF) energy, electron beams or other accelerators pumping the medium, optical excitation from light sources, nuclear reactions and expanding a hot gas into a vacuum (gas dynamic excitation).

In each case, when the proper conditions exist, the result is laser light. Sometimes it will be visible, and other times the wavelengths will be in the infrared or ultraviolet regions. But, by any other name, it is still a laser. My apologies to,...uh, Shakespeare? Well...somebody.

Anyway, let's take each of these alternative methods and see what they have produced in terms of working systems. At the same time, I hope to familiarize you with a number of commercially prominent laser types.

Chemical excitation has yet to make a big splash, as its application results in some rather complex schemes. The most commonly seen lasers from this category employ the hydrogen halide group of chemicals. These include hydrogen fluoride, hydrogen chloride and hydrogen bromide in a gas form.

All are both toxic and caustic in nature, which leads to a special set of problems. For example, hydrogen fluoride will dissolve glass; hence if you're going to make a laser using this gas, you don't want to make it out of glass. Trust me, you will soon regret such a decision. Not only will your laser melt down, but you might very well poison yourself to death. (Hydrogen fluoride is nasty stuff!)

Additionally, the mechanics of such a system are, well...shall we say, less than user friendly. The gas has to be fed into the laser at a critical rate and removed (vented) rapidly to prevent a collapse of the population inversion.

To date, this type of device has found little attention outside the scientific community. There, they do play a significant role in spectroscopy and chemical studies. Beyond that, the only other interest has been in developing high-power systems for the military.

Next on the list is the *RF excited devices*. Actually, this is more a *process* than a specific laser type. By exposing the gas to radio frequencies, often in the microwave range, a population inversion can *sometimes* be achieved. Sometimes, however, it isn't achieved, and that makes this approach somewhat iffy at best.

So, don't get too enthusiastic about constructing an RF laser. You might not accomplish that goal. And, even if you do, the energy source is going to be high powered and expensive. I merely mention this form of excitation for your knowledge rather than as a practical fabrication strategy.

A variation of discharge excitation, the *electron beam method* also applies energy to the gas with electricity. But this is done with some form of electron accelerator (such as an electron gun) instead of a discharge between a cathode and anode.

Again, these lasers have found little use outside the laboratory. While they are fast, they require cumbersome and expensive equipment to produce the electron beam.

Optical excitation for gas lasers is not nearly as common as it is with solid-state systems, due largely to low efficiency. In fact, this efficiency level is so pathetic, these lasers are really only good for producing specific wavelengths. Hence, they also reside mainly in the laboratory.

There though, they do serve a useful purpose for laser demonstrations and other scientific research. While this is an infrequent technique for gas excitation, it is worth being aware of. Who knows, it might be the *Final Jeopardy* question when you appear on that TV show in all your glory. Well, maybe not. But it could happen! Couldn't it?

Alright, enough of that! Let's move on. *Nuclear excitation*. This is a neat one. As you might expect, however, it is not used a great deal. But the principle is the same. Derive a population inversion by applying an appropriate energy level to the medium.

In reality, this works in a two-state fashion. The nuclear reaction produces powerful pulses that result in isotopic fusion. The gas medium is then excited by ions produced from this fusion process. This results in some very high outputs.

So high, these lasers were considered for the Strategic Defense Initiative (SDI) program during the Reagan administration (mid-1980s). Work in this field goes back ten years earlier or so, but as of this writing, a functional laser has yet to hit the private sector...and, may never hit the private sector.

Our *last category* involves a novel approach in which gas under high temperature and pressure is released into a vacuum chamber. As the gas expands, it cools but still retains enough energy to create the needed population inversion.

Due to this cooling, the process is quick, and like the chemical excitation systems, the gas flow must be constant to maintain the inversion. This, of course, adds significantly to the scientific complexity and mechanical intricacies of this form of light amplification by stimulated emission of radiation (LASER).

They can produce very high outputs, and thus have a value in the laboratory. Although other varieties have been built, this type of laser is usually classified as a *carbon dioxide*, and commercial versions are not yet available.

With that discussion under our belts, let me get a little more specific in terms of actual lasers. The following will be a short description of many of the common lasers available to both the scientific community and commercial market. **In some cases, I will even include a picture**!

CARBON DIOXIDE LASERS

Easily the most versatile of high-power gas systems, the carbon dioxide (CO_2) laser dates back to the early years of laser research. While CO_2 devices come in a number of flavors, the one most often seen is the conventional discharge sealed tube (see *Figure 11-1*).

This construction is similar to a helium-neon system in that it contains a gas mixture in a glass envelope and utilizes both fully and partially reflective mirrors. However, instead of the more intricate bore/spider arrangement of the HeNe tube, the carbon dioxide enclosure uses small electrodes near each end.

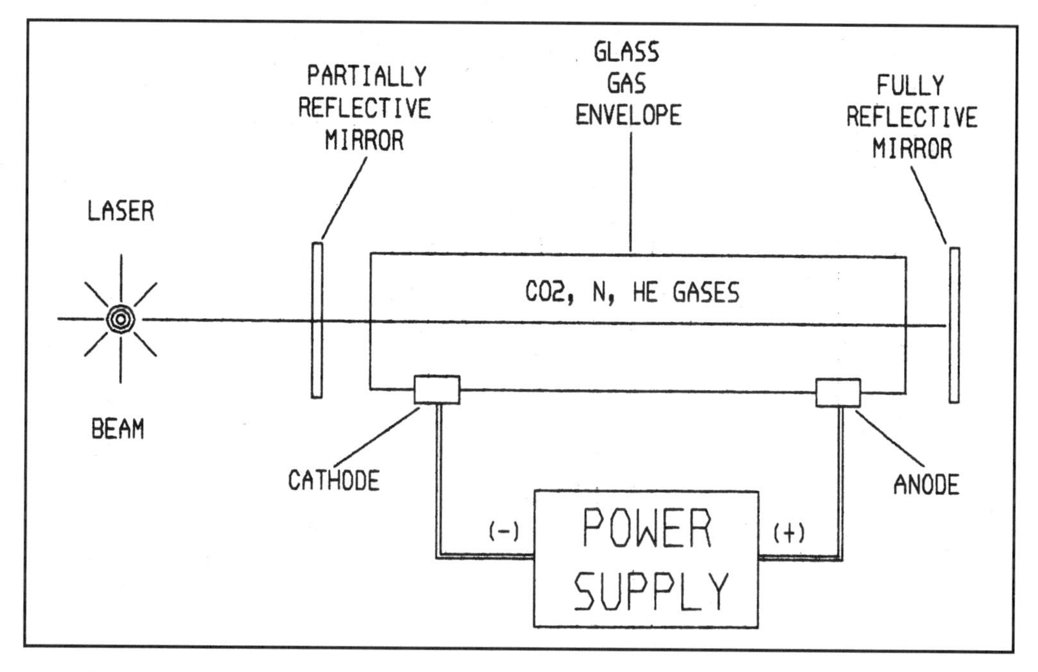

Figure 11-1. The carbon dioxide gas laser. This is probably the most common configuration. However, there are several other less common types.

A direct current (DC) potential, usually lower voltage but higher amperage than a HeNe, is applied to the electrodes, and the result is an infrared beam emitting from the partially reflective mirror. However, this output can range from 1 watt to several hundred kilowatts, making this device one of the most powerful on the commercial market.

The gas used is a combination of carbon dioxide (CO_2), nitrogen (N) and helium (He). The CO_2 generates the light, while the nitrogen helps the excitation process. Helium acts as a buffer that helps manage the substantial heat developed and also aids in maintaining the population inversion.

Surprisingly, the gas mixture ratio is usually somewhere in the 10 percent CO_2, 10 percent N and 80 percent He range. This, of course, will vary with different manufacturers, but the total amount of CO_2 is always going to be substantially lower than the helium. I wonder why they call it a carbon dioxide laser?

Oh well, they do, and it is a winner. These gems are used for everything from delicate eye surgery to cutting steel plate. And, remarkably, are relatively safe to use as long as certain precautions are observed.

ARGON AND KRYPTON LASERS

Also known as *noble gas ion lasers*, these represent some of the most important, most germane and most dangerous devices in the world of commercial lasers. All should be handled with extreme care!

Since both systems operate in virtually the same fashion, I will discuss the argon laser. The only difference between the two is the gas used, and in some cases both gases are mixed to produce two separate wavelengths.

Actually, like most gas systems, argon lasers emit light in more than one wavelength, but the strongest line usually will be around 514 nanometers, or a greenish-blue color. Krypton's strongest wavelength is usually 641 nanometers, or red, although it can produce yellow, green and violet as well.

As a sidebar, two other noble gas ion lasers are common, namely, the *neon ion* and *xenon ion* varieties, with wavelengths in the ultraviolet and blue-green respectively. Both are minor players, but again, it is nice to know about them.

Anyway, the structure of an argon device is quite complicated and very temperamental (see *Figure 11-2*). While it may not appear so from the diagram, the combination of water cooling, gas replenishment and high current demands makes this a tough customer to design and build.

Current from a separate power supply flows from the cathode to anode along the bore. Due to the intense heat generated, the bore is normally constructed from beryllium oxide (BeO) ceramic which can handle the temperature. Both bore ends are sealed with optical openings that are coupled to external resonant mirrors and employ *Brewster*, or angled, windows.

To dissipate the enormous amount of heat, a water jacket surrounds the bore to sink the heat. Naturally, this provides complications regarding keeping the water where it is supposed to be. Anytime—I mean *anytime*—you have to incorporate water flow into a design like this, there is going to be trouble.

Also surrounding the bore, outside the water jacket, is a magnet that helps concentrate the current flow in the bore. This is not always used, especially in low-power devices, but does help improve the rather poor efficiency of argon lasers.

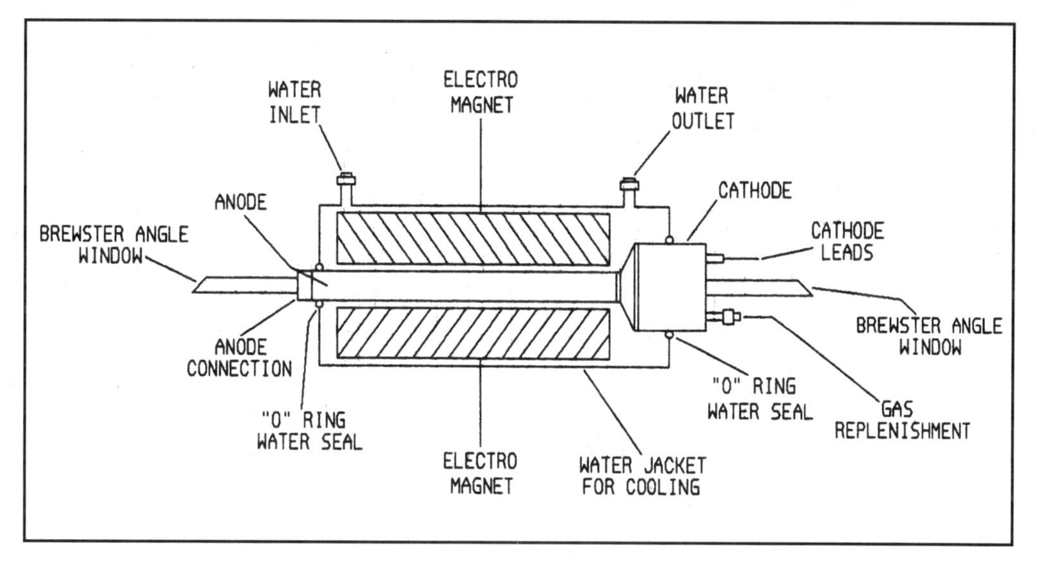

Figure 11-2. A "generic" diagram of an argon gas laser. This system is water cooled, with electromagnets (for beam concentration) and Brewster windows.

The gas, which has to be constantly replenished, is fed from a reservoir to the bore interior through a gas fill tube. Here again, leakage problems are a natural with such an arrangement. If the gas is not renewed, contamination and gas depletion will often lead to tube sputtering.

So, as can be seen, ion gas lasers are a double handful in terms of practical construction and operation. They do serve valuable purposes in both the scientific arena and in commercial applications such as fingerprint identification.

NITROGEN LASERS

One of the early gas lasers (1963), the nitrogen variety is another discharge device. It is also one of the simplest lasers to build. So simple, it doesn't even require resonant mirrors, although they will improve the overall output.

Nitrogen lasers have been featured in more than one amateur scientific magazine, and are a big hit for science fair projects. All you need is a glass tube with a cathode at one end and anode at the other, a source of nitrogen gas and a rather high-voltage power supply (ranging from 10 to 40 kilovolts). This last one, of course, presents the biggest problem.

You could add a fully reflective mirror and another partially reflective (about 5% reflection) mirror, but these are not necessary for proper operation. In fact, the reflective surfaces will probably not improve the device's output level by more than double.

The ultraviolet wavelength is very precise and is not a tremendous eye hazard unless your are dealing with a high-power nitrogen laser. Even so, it's never a good idea to let laser light of any sort enter your eyes, no matter what the wavelength.

The major hazard connected with these systems is the voltage/amperage from the power supply. Both are way up there compared to the likes of a helium-neon laser and can be fatal. Hence, extreme care must be taken when working with nitrogen lasers to prevent shock and/or death.

Due to their vary narrow wavelength, these devices are found in many laboratory settings. From chemistry to biology to optics, the nitrogen laser is held in high esteem by scientists and hobbyists alike.

HELIUM CADMIUM LASERS

I personally consider this device a gas laser, although I occasionally get an argument on that classification. Since the system is based on ionized metal vapors, it seems logical to me that it be designated a *gas* medium. But, the contention has been made that cadmium is a solid and therefore the structure should be considered *gas/solid-state*. But, I will stick to my guns due to the fact that both materials are in a gaseous state. Plus, it's my book and I can do that sort of thing. So there!

Anyway, it doesn't really matter all that much because this device does work quite well. As one of the first metal vapor specimens, HeCd lasers can produce power in the hundred milliwatt range on the 442 nanometer, or blue, wavelength. They can also produce up to 50 milliwatts of ultraviolet (around 325 nM), and this fact makes them very appealing for many applications.

Regarding design, the helium-cadmium systems are similar to most other discharge style lasers. Research revealed that cadmium is an excellent laser medium, and the addition of helium furnished better efficiency and continuous operation.

Referring to *Figure 11-3*, it is seen that a typical helium-cadmium device is shaped somewhat like a helium-neon tube with a few additions. (There is an older design that has separate sections for the helium and cathode element, but for our purposes, we will stick with the newer scheme.) The laser utilizes the elongated glass envelope, but a fluorescent lamp acts as the cathode, and a heating element keeps the cadmium in the vapor state.

Additionally, the outer sections handle the helium and cadmium supplies. Since both are depleted during operation, it is necessary to provide a reserve of each. The amount of reserve largely determines the life expectancy of the tube (usually in the 6 to 10 thousand hour range).

The bore is suspended in the center of the device, and like the HeNe tubes, is a major determining factor concerning beam size and power. A separate smaller bore acts as the anode and also helps establish a condensing chamber necessary to collect the cadmium as it cools and returns to a solid state.

The ends of the tube are hard sealed to the glass envelope with alignment flanges, and as might be expected, the partial and full reflective mirrors are located in these assemblies. The mirrors, of course, provide the needed resonance to achieve and maintain the population inversion.

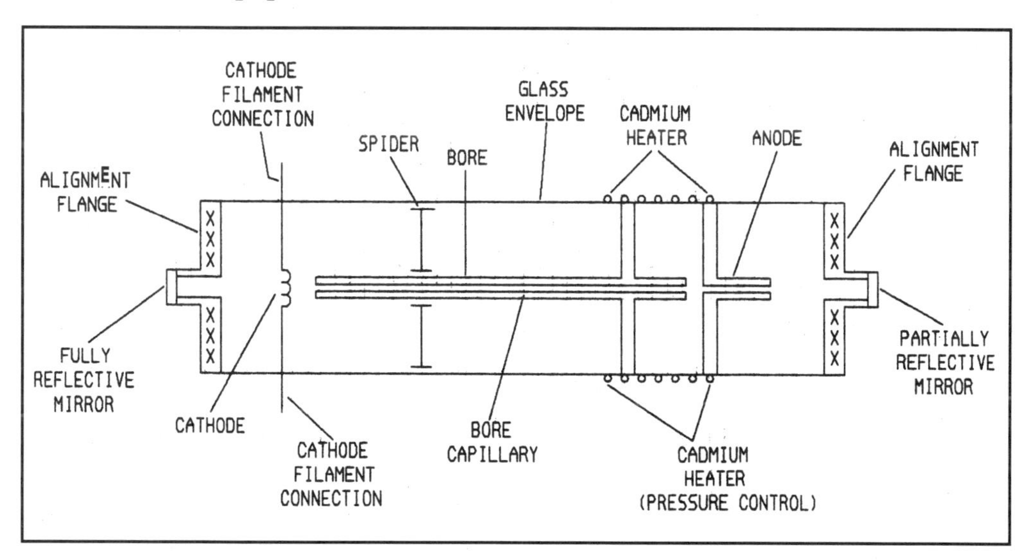

Figure 11-3. A typical helium-cadmium laser. The design is similar to the HeNe tube. There is an earlier segmented configuration that is more complex.

Due to the heating element and operational heat generated by the systems, an airflow cooling scheme is necessary. And, very precise helium/cadmium concentrations and pressures must be observed to prevent failure of the laser.

However, all things considered, the helium-cadmium device is a relatively inexpensive and simple laser system. Hence, it has garnered a good deal of attention both in the scientific community and the commercial arena.

COPPER/GOLD VAPOR LASERS

These devices, known as *neutral metal vapor* lasers, are similar in construction to the ionized metal vapor system, but they function in a different fashion. First, the only gas present (other than a small quantity of neon for improving discharge) is the metal vapor itself, and the electrical discharge excites the atoms of copper or gold to produce the inversion.

They are usually linear in design, but due to the high temperatures needed to maintain the metal vapor (1500 to 1850 degrees Celsius) the enclosure cannot be made of glass. The material of choice is usually ceramic alumina shaped in tubes.

One beneficial discovery occurred in the Soviet Union in the early 1970s. This involved using the natural heat generated by the laser action to help heat the system. This approach has been adopted by many manufacturers and has increased efficiency.

Today, only the copper and gold vapor lasers have proven to be commercially prudent. Copper systems can provide power in the 100 watt area, in the yellow and green wavelengths, while gold will render several watts in the red region. This does lend appeal to both types of devices.

As with other metal vapor designs, replenishment of the metal is essential to prolonged operation. This is due partially to the fact that the copper or gold vapor will condense into its metallic forms on cooler surfaces within the system; thus, additional metal must be introduced.

Like the nitrogen laser, these devices do not necessarily require resonance mirrors. But again, like the nitrogen systems, mirrors will improve their already high efficiency rating. With many designs, 10-percent reflectivity of the partial mirror is more than enough to maximize the system performance.

These laser types, especially the copper vapor units, are quite dangerous. Both species emit visible light that can easily do irreparable damage to the retina, possibly resulting in blindness. And, the high voltage/current rating of the power supply is nothing short of LETHAL! On a last note, these systems produce very high temperatures and will burn the fire out of you if you don't let them cool down properly. Excuse the pun!

While fairly simple to construct, the materials and support systems make the neutral metal vapor lasers hardly the best choice for experimenters. They tend to end up being quite expensive. Additionally, the safety hazards associated with these designs demand very cautious handling. Again, they are not a good selection for a home laboratory laser.

But, commercially and scientifically, these are true winners. Due to the wide range of wavelengths and relatively high power, they have found their way into the hearts of more than one engineer and/or scientist.

XENON/XENON-HELIUM LASERS

Now we are starting to get down to some of the more obscure, yet still important, gas lasers. While these are not meant for "home-brew," they do serve a viable role in the research arena.

The first of these is the xenon (Xe) laser. Theses are pulsed systems that operate in the blue-green region with powers as high a one kilowatt. With most devices, the predominate wavelengths will be in the green at 526 to 540 nanometers.

They are true pulse lasers, but the pulse rate can be as high 200 hertz. Thus, to the human eye/brain, they will appear to be continuous in operation (good ol' *persistence of vision*). It is this attribute that makes them attractive to various manufacturing interests.

Like any high-power visible light laser, xenon units are extremely dangerous to work with. All appropriate, and maybe some not-so-appropriate, precautions must be taken to prevent serious damage and/or possible death. If you encounter these systems, do not take chances with them!

As of this writing, the electronics industry is probably the single most important employer of xenon lasers. The power and short wavelengths of Xe devices make them ideal for the highly precise standards involved in microcircuit fabrication. Aside from that, they have been utilized in holography, lithography and research requiring relatively short wavelengths.

Our second variety is the *xenon-helium* laser. This type has garnered far less attention than the plain xenon version, but has found a home in the laboratory. Or, at least, some laboratories, anyway.

Xenon-helium systems operate in the near infrared region and function in a relatively high power mode (about 200 milliwatts)of operating pulse. They can, however, demonstrate continuous wave performance, making them useful research tools.

Structure is often linear with flowing gas for the commercial models. But, low-power units can be closed, or *sealed*, and have reasonably good life expectancies.

NITROUS OXIDE LASERS

So similar to carbon dioxide/carbon monoxide systems, the CO_2 and CO gases are often replaced with the nitrous oxide (N_2O) gas mixture to produce this type of device. These lasers manifest very precise far-infrared wavelengths. The most frequently encountered gas combination is 12 percent N_2O, 16 percent N, 9 percent CO and 63 percent He.

Power levels of 10 watts continuous and 30 watts pulse are common for nitrous oxide devices, but this is substantially lower than CO_2 systems. However, when a very specific wavelength is required, this is your boy! And in that lies the secret of the moderate success N_2O lasers have enjoyed.

The gases are usually introduced at one end of the system, passed on through, then removed at the other end. This procedure works well, but due to the toxicity of the mixture, extreme care must be taken when disposing of the used gas.

As with CO_2 devices, the power supply is nasty and can fry you. So pay attention if you have to deal with said supply. Trust me, you will look quite funny with very straight hair that points north. And that could be the very least of your worries if you should tangle with this amount of power.

To date, spectroscopy and far-infrared laser pumping are the primary application of nitrous oxide devices. Lower power output, in contrast to CO_2 systems, has prevented any substantial commercial application of these lasers, but that may well change in the future.

CARBON MONOXIDE LASERS

Most carbon monoxide (CO) systems are modified carbon dioxide lasers and are capable of both continuous and pulse operation. They usually function in the mid-infrared range (5 to 6 micrometers), which makes them ideal for spectroscopy. Another area of interest has been the medical field, as infrared lasers have proven highly proficient in this arena.

Like their CO_2 cousins, carbon monoxide devices are electrical discharge in nature. With proper cooling, output power can reach the 8 to 10 watt range due to the high efficiency they exhibit (up to about 60 percent), and many researchers feel the systems have the capability of extremely high power performance.

Such sentiments led to exploration into using CO lasers for military weapons programs. However, unusually high atmospheric absorption of their wavelengths made them unsuitable for any type of ground-to-outer space application. Hence, one highly promising employment opportunity was lost, resulting in a meager deterrent effect on further research and development.

The gas mixture for these lasers is usually carbon monoxide, helium, nitrogen, xenon and sometimes air. It is pressurized in both the continuous and pulse models, and almost always applied in a *flowing gas* fashion.

As with the nitrous oxide systems, the high toxicity of carbon monoxide dictates extremely careful disposal of depleted gas. And, the high voltage/current potentials of the power supply require caution when operating such lasers. Additionally, just about all CO devices demand some form of cooling to maintain proper efficiency and prevent damage to the system. This can range from air cooling with lower-power units to cryogenic chilling for high-power systems.

IODINE LASERS

This last gas laser, the iodine (I), is an unusual one. It sees a majority of service in the research laboratory, but iodine lasers have also caught the attention of military weapons programs.

There are two basic types of iodine devices: the oxygen-iodine and photodissociation iodine systems. Each has its own purposes, but both produce a near infrared line in the 1,300 nanometer range.

With the *oxygen-iodine* units, oxygen is produced through a chemical reaction and then transfers its energy to the iodine. This energy transfer produces the needed population inversion. *Photodissociation* utilizes flash lamps to provide the necessary pumping and can generate outputs in the tetrawatt range.

While normally pulsed lasers, the iodine variety can be run continuously at much lower power outputs (30 to 40 milliwatts). These smaller devices are good sources of near infrared for experimentation, while the high-power devils have numerous applications including the aforementioned laser weapons.

LIQUID LASERS

Over the years, a number of liquid lasers have been theorized and/or tested, but the only one that has endured is the *dye laser*, so for that reason I will confine this section to them. It will be worth the effort, however, as these are very interesting and versatile systems.

Officially called the *organic dye laser*, these devices have found a welcome home in the scientific community. Dye lasers can be tuned over the entire visible light range, and, through the application of special crystals that double the frequency of light waves, the range can be extended well beyond visible light.

Naturally, being able to set the output wavelength thrills anybody involved in spectroscopy, optics or photochemistry, not to mention a number of other scientific fields; hence their laboratory popularity.

However, these devices are not simple! They are, in fact, both complicated and temperamental in structure and operation. This, of course, makes them expensive—to manufacture as well as maintain. The nature and reason for these drawbacks will become more apparent as we get into this discussion.

However, when a scenario calls for these systems, expense often plays second fiddle in the requisite chain of priorities. A dye laser's ability to produce extremely short pulses and very narrow wavelength lines qualify them as the only practical light source for a particular job. In such situations, researchers frequently consider the system cost acceptable if the end result is justified.

To attempt to define the structure of dye lasers in general is a somewhat futile effort. There are literally dozens of different configurations. So, in light of that, I will instead describe the basic requirements of these devices.

One of the key elements that make dye lasers popular, and financially feasible, is their talent for tuning (changing) the output wavelength. All dye lasers bear this ability, and it can be accomplished in several ways.

The trick to this process is isolating the wavelength you want to emit. Naturally, this is partially the job of the dye, but additional help comes from the illuminating source and optical system.

Many dye lasers utilize either diffraction gratings or prisms to segregate the wavelengths. With a diffraction grating, some of the *pump* light is channeled through an optical assembly to the grating, and by adjusting the angle of said grating, the reflected light emerges at the proper wavelength.

When using a prism, some of the pump source is again channeled through the prism (which remains stationary) to the rear cavity mirror. The mirror is then adjusted, permitting the light beam to return to the prism at the right angle. This results in the correct color being reflected back into the *dye cell.*

Speaking of the dye cell, this is a container that holds the selected dye, or can be a jet arrangement that sprays the dye into the light path. Either way, the specific dye employed plays an important role in the final emitted wavelength, as the dyes filter white light, only allowing the intended color to emerge.

Some dye lasers utilize *flowing dye* systems instead of a sealed vessel. There are a couple of benefits to this approach. By keeping the dye moving, it stays uniform, and the arrangement allows for easier dye replenishment.

As with most lasers, mirrors and a bore are employed to direct the laser beam. The mirrors are of the usual type (fully and partially reflective), and the bore's size largely determines the diameter of the beam. This should sound familiar.

One difference concerns the type of *pump* light source. These can vary from xenon flash tubes to external laser systems, depending on what form of output (continuous/pulse) and power rating you want the system to deliver.

With a flash tube arrangement, high-output tubes, usually helical, are employed, and the light is often amplified with optics. That light is then introduced to the dye cell and other related optical elements.

When an external laser becomes the pump source, the dye cell is the target of the beam. Again, additional optics are normally implemented to help restrict the wavelength. One advantage of a laser pump is better control over the light intensity. Also, high-power lasers can produce high-power dye systems.

All in all, being able to choose a pumping source contributes a certain flexibility to dye laser systems. It allows the operator to match the best light source to the task, and that can be beneficial both economically and pragmatically.

Let's talk about the dyes for a moment. They are special *fluorescent* organic chemicals dissolved in some sort of solvent. A solvent can be anything that allows the dye to dissolve in it and still maintain the proper characteristics that permit lasing action. Water would meet this criteria except that a vast majority of fluorescent dyes will not dissolve in it. So, organic chemical solvents are most often employed.

As mentioned earlier, dye replenishment is necessary with just about all dye lasers. This is due to dye degradation from the pumping source and heat of the system. The stronger the light applied to the dye the faster it will decompose. At least that is a general rule of thumb. Some dyes are more resilient than others.

Also, ultraviolet light will usually tear up the dyes much faster than other wavelengths. Ample research has been directed towards developing dyes that have better resistance to these destructive elements. But, due largely to the nature and structure of dye molecules, this has proven a difficult task.

Furthermore, the solvent and other additives play some part in the longevity of the dyes. Again, extensive exploration into better combinations of dyes, solvents, et cetera, has been helpful, but dye deterioration remains a problem.

In the way of safety, dye lasers probably have more hazards to contend with than any other type (barring the nuclear pumped x-ray jobs). In addition to the usual peril to your vision and menace of shock from the high voltages present, these systems can poison you with the dyes and solvents.

Most of the chemicals used are at least somewhat toxic and can be absorbed through the skin. Thus, you do not want to get them on you. And, if you do, wash them off thoroughly and immediately.

Additionally, many of the solvents are flammable. Hence, they will burn—as in a spark from the power supply or flash tube setting the whole damn thing on fire. Not a good deal! Consequently, when using these devices, take great care not to expose them to an open flame or other source of fire!

In conclusion, the dye laser is not exactly a hobbyist's project. They are expensive, problematic and fairly dangerous. However, they do hold an important place in the world of the laser. If you have any doubts, talk to the folks in the scientific domain. They will reassure you as to the dye laser's value.

SOLID-STATE LASERS

Here we have a classification of lasers that is quite diverse. It includes all devices that do not employ liquid or gas as their medium; in other words, all solid materials that display laser action. All except the semiconductor, or diode lasers. They will be covered next.

As you probably remember, the first working laser was a solid-state device utilizing a rod of synthetic ruby sapphire as its medium. This task was accomplished by Theodore Maiman of the Hughes Research Lab who, while others debated how a laser should be constructed, simply built his. And it worked!

This was indeed an historic event, but aside from presenting the world with the first working laser, it also established the cosmos of solid-state designs. And that is what we are going to talk about now.

A majority of this category consists of a rod made from either glass or synthetic crystal, a flash lamp and resonance mirrors. They also use reflective, or optical, cavities that surround the rod/flash lamp assembly and help direct the light to the rod.

In addition to the flash lamp approach, most solid-state devices can also be pumped with other lasers. Semiconductor units are a favorite, but nitrogen and carbon dioxide have been known to serve this purpose.

So, with all that said, let's take a gander at some of the more important solid-state lasers. This is not to say the others aren't also important, but there is a limit to how long I can make this chapter.

RUBY LASERS

Interestingly, the first laser type is still with us. Unlike some that have fallen by the wayside, the ruby rod just pulses along. Albeit they aren't quit as popular as they once were. However, such longevity is admirable by any standard, let alone the fickle world of lasers.

OK, the medium (rod) is a crystal grown from synthetic sapphire with about half a per cent of chromium impurity. The chromium is what gives the rod its red color and generates the lasing action.

Ruby lasers are pulse lasers with a wavelength of 694.3 nanometers. That translates to a deep red flash of light each time the pump lamp fires. They are also capable of easily burning holes in a variety of otherwise sturdy materials. Incidentally, unlike most members of the solid-state family, ruby units cannot be pumped by other lasers. And, for your enjoyment, *Figure 11-4* illustrates a typical ruby rod device.

They do experience problems with overheating, and some models have to be water cooled. Generally, unless the rod is dirty or overly stressed, ruby lasers will not be damaged by this heat, but their performance will. Thus, some form of cooling, even if it is just forced air flow, is strongly advised.

The mirrors are of the same flavor as most other lasers; that is, one fully reflective and one other partially reflective. However, the coating is applied directly to the ends of the rod, as opposed to a separate surface. Naturally, the principle is identical—a resonance system that promotes the population inversion.

As with most solid-state systems, a ruby laser needs little maintenance. The flash tubes have to be replaced from time to time, and the system needs to be kept clean. Otherwise, unless you experience an electrical/electronic problem, the laser will function flawlessly.

Regarding safety, this can be a treacherous piece of equipment. A single flash to the eye can result in permanent retinal damage and even blindness. Additionally, the flash lamps require lethal levels of power.

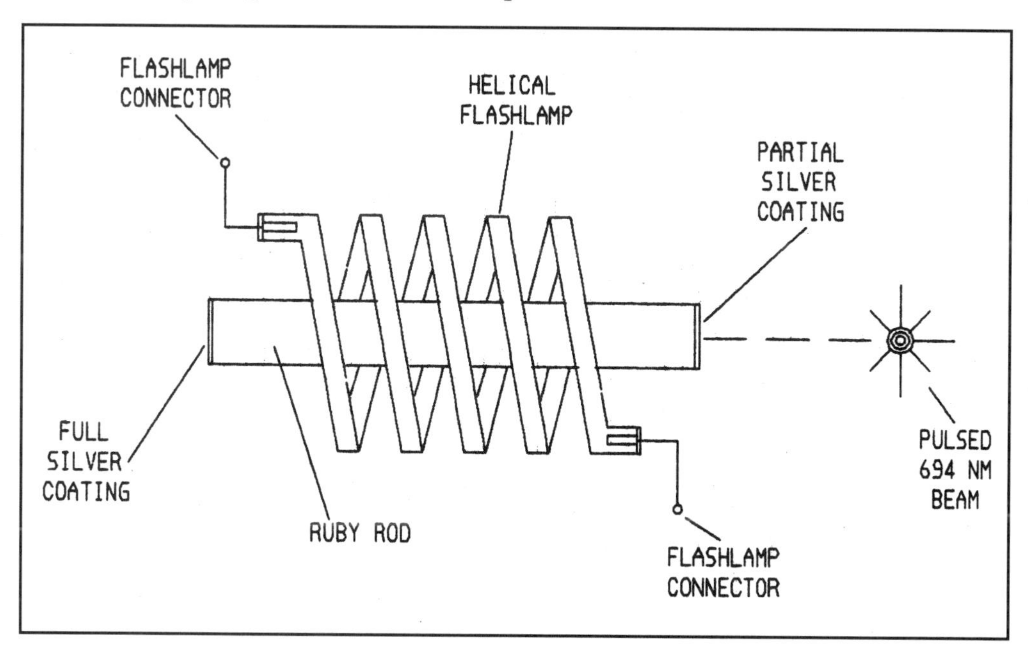

Figure 11-4. The ruby rod (solid-state) laser. This is the classic design. There are several other configurations that use multiple linear flash lamps, etc., but this was the first.

Hence, when the system is operational, its electronics, especially the capacitor banks, should be avoided at all cost. For that matter, those capacitors can hold a dangerous charge long after the laser has been turned off. So, if any exposure to the flash lamp circuitry is anticipated, discharge the caps.

NEODYMIUM LASERS

The neodymium group is by far the most popular and versatile of the solid-state lasers. I say *group* because there are a multitude of variations employed with these lasers, and I will discuss the most significant of those here.

Neodymium is one of the *rare earth* metals and is the principal active element in these devices. Actually, it is *triple ionized neodymium* in a glass or crystalline host. But, the important point is that the external energy source excites the neodymium, which produces the population inversion.

Said external energy is always *light* with these lasers. This can be an arc lamp for continuous operation, a flash lamp for pulsed performance or a semiconductor laser for either. Naturally, the intensity of this source plays a major role in the power output of the laser.

And, that output can be quite high. The average power of short-pulse systems can be in the kilowatt range, in the near infrared region, with peak power in the gigawatt region. Continuous devices, however, usually operate in the milliwatt realm, with the exception being a few high complexity devices.

As mentioned, the host material can be either crystal or glass, and both have their own characteristics and advantages. Conversely, they also have their own handicaps. So in the next section I will discuss and compare these two mediums.

First, though, let me cover one similarity. In both cases, the neodymium is an impurity. This impurity is introduced by a process called *doping* which merely means the neodymium is included in the host as trace amounts (usually about 1 percent).

GLASS RODS

In glass, the impurity is simply mixed into the molten mass as you would mix in chemicals to color the glass. It is, of course, evenly dispersed throughout the laser rod. This process is relatively rudimentary compared to the crystal approach.

Glass also allows for much larger rods to be constructed and is substantially cheaper than crystals. These factors have contributed to glass's popularity and some very large and powerful lasers. Lasers such as the "Nova" device at the Lawrence Livermore National Laboratory.

Glass, on the other hand, suffers from poor thermal dissipation when compared to crystal materials such as YAG and YLF. Hence, the rate at which the pulses can be delivered has to be decreased. The Nova laser can only be fired about once an hour.

As a note, the plight related to poor thermal dissipation is more a matter of degraded optical quality than a danger of melting or shattering the rod. Quality of the emitted beam will suffer substantially unless the rod is kept reasonably cool.

CRYSTAL RODS

The crystal versions are usually based on one of two synthetic materials: yttrium aluminum garnet (YAG) and yttrium lithium fluoride (YLF). Both have a garnet-like structure and are relatively brittle in nature, although YLF is less rigid and hard than YAG.

Both demonstrate much better thermal characteristics than glass, allowing them to be used at room temperatures (no artificial cooling), even in continuous operation. Also, the optical properties of these substances is superior to that of glass. That, naturally, has an influence on the quality of the beam.

Crystals, however, have to be grown under laboratory conditions, and the process is difficult and slow. It is also expensive relative to glass. Additionally, the neodymium has to be incorporated as part of the crystal's chemical structure, which adds further complications.

So, as can be seen, each medium has its place and purpose in this realm of laser science. When large high-power devices are in order, glass is usually the best choice. But when super quality is crucial, crystal hosts become the requisite selection.

One advantage to these systems is that amplification is a moderately simple task. Chunks or disks of either glass or crystal introduced into the beam path provide very efficient amplifiers. This aspect has contributed to lasers of exceptionally high output.

As with many other solid-state systems, mirrors are used to generate a resonance mechanism. However, with all but the very smallest units, these mirrors are external. This also allows other optical elements to be placed within the resonance substructure.

When arc or flash lamps are employed for the pumping, a reflective shield, or *cavity*, is used to help direct as much energy as possible to the rod. This is similar to the arrangement utilized with ruby rod lasers.

If, however, a semiconductor laser acts as the pumping source, its beam is usually directed at one end of the rod. This setup eliminates the need for a fully reflective mirror on that end, as the diode laser provides ample energy to keep the inversion alive. *Figure 11-5* illustrates a typical *semiconductor laser* pumped system.

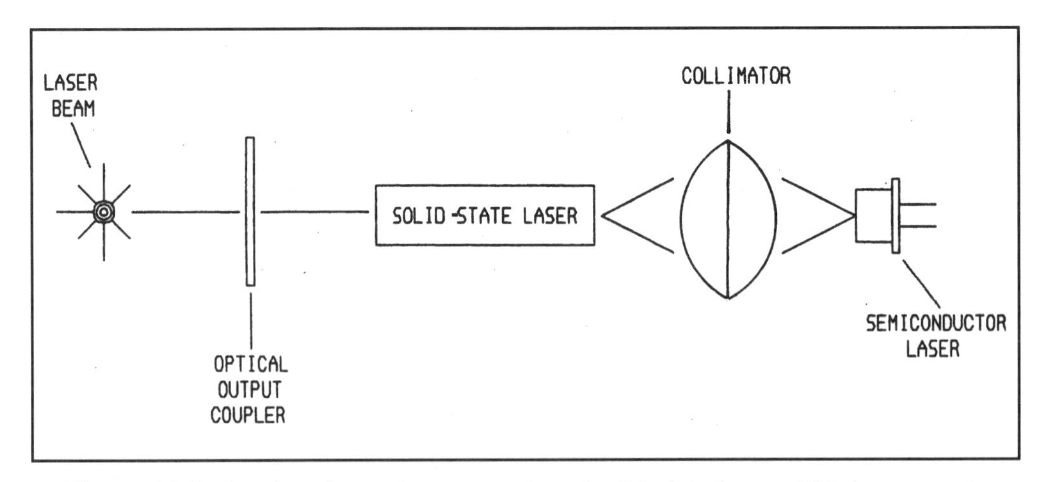

Figure 11-5. Semiconductor laser pumping of solid-state lasers. This is a generic diagram of a traditional method to pump a solid-state laser. Usually the diode laser is in the infrared range.

That gives you an overview of the neodymium lasers. Over the years, a multitude of different designs and variations on those designs have come and gone. And, many have remained with us. But, if there is a single truth in all this it is that neodymium lasers are here to stay.

TUNABLE SOLID-STATE LASERS (VIBRONIC)

Using a process similar to dye lasers, some solid-state devices are able to emit many different wavelengths. Not all at the same time of course. These systems employ internal atomic transitions (vibrational states) to *tune* the output, and are nearly as versatile as their dye counterparts.

I say *nearly*, as they still cannot produce the entire spectrum of the dye devices. Usually, the range is from near infrared to near ultraviolet, which tends to exclude the true ultraviolet/infrared emissions of dye lasers.

Again, synthetic crystal material, often sapphire, is doped with impurities (titanium, chromium, etc.) that provide the active ingredient for population inversion. Some research into using glass has been done, but the results are generally not as satisfactory as with synthetic crystals.

The structure of most tunable solid-state lasers is again similar to other types. Essentially, the rod is pumped by an external light source, almost always a semiconductor laser, and a mirror/optical arrangement generates the beam. That is a somewhat simplistic description, but I trust by now you get the idea.

With the introduction of the titanium-sapphire (Ti-sapphire) variety, it seemed, at least at first, that the best of two worlds had been achieved. However, the true nature of the beast (with its wavelength shortcomings) soon reared its ugly head.

Since the excited state *period* of most tunable solid-state lasers is considerably shorter than YAGs, YLFs and so forth, these devices are best pumped with semiconductor lasers. Flash lamps are simply too slow to properly pump this type of device; hence, short pulse or continuous lasers are used.

Many different synthetic materials have proven satisfactory for this variety of device: alexandrite, forsterite, emerald and thulium YAG, just to name a few. Each has its own demands, but all have proven effective in this area.

Research continues in this relatively new field, and the promise of better vibronic lasers looms in the future. If the bandwidth and short excited state obstacles can be licked, these systems may well replace the more complex dye lasers.

A FEW OTHER SOLID-STATE LASERS WORTH MENTIONING

In this last SHORT section, let me introduce you to several other solid-state devices of less prominence than the ones we have reviewed. They all work in the near infrared or infrared region, and all are structured in the usual manner.

These include *thulium*, *erbium*, *crystal erbium*, *holmium* and *color-center* lasers, with all but the last type employing rare earth metals. In the first four, trace amounts of rare earths act as the active element and is excited to a population inversion.

The last specie is another specialized tunable wavelength (vibronic) device doped with a variety of impurities. Each different "contaminant" determines the *center* of the *color* band the laser will produce, hence the name *color-center*.

Each of these devices is rather new to the laser world, and research continues into even newer and better systems. As novel designs emerge, they will undoubtedly be added to the already massive number of existing lasers.

SEMICONDUCTOR LASERS

Of all the major laser categories, this one has to be considered the newest, even though the concept of a semiconductor, or diode laser closely followed the invention of the light-emitting diode (LED). But, due to complications involving structure, operating temperature and the like, it has only been recently that semiconductor devices have proven viable.

Thus, I consider this area the newest. It is also one of the most prolific in terms of innovation and application. Today, there are probably more laser diodes in use than all other laser types combined. And, due to reliability and cost, that number can only increase.

I have discussed semiconductor devices at length in other chapters, so rather than go back through all the theory and structure, let me acquaint you with a couple of the more curious—or should I say bizarre—members of this gang.

One recent addition that caught my eye is the high-power infrared jobs. These are packaged in a standard TO-3 case and can produce outputs from 1/2 to 1 watt. They do emit infrared, but with the addition of a frequency-doubling crystal, the wavelength can be changed to the green and blue regions.

These high-power devices are the product of an incredible technology known as *quantum cavity* construction. In short, the junction area of these lasers is so small that it is often only a few hundred atoms wide. At such a minuscule size, efficiency is extremely high, and virtually all the energy emits as a beam.

Here again, continuing research promises to produce even better designs. All this translates into higher power outputs and hopefully a greater variety of wavelengths.

The second type is affectionately referred to as the *lead salt* semiconductor lasers. Essentially, all of these use lead in some form and emit in the far infrared spectrum (3.3 to 29 micrometers). These factors have somewhat restrained interest in the devices, as lead salts can be problematic and those wavelengths are not especially appealing outside the lab.

However, they have displayed some useful, if not valuable, traits that do make them popular in certain scientific areas. Among these fields are high-resolution spectroscopy, fiber optic communications and infrared detector testing.

The lead salts are frequently compounds of lead with tin, selenium, tellurium, sulfur, and the rare earth europium comprising the rest of the molecule. All produce highly crystalline structures, which accounts for the wavelengths.

Lead salt lasers are also temperature sensitive; that is to say that output wavelength varies with temperature changes. When dealing in shorter wavelengths, this is less than desirable, but in this spectrum, it is actually an advantage. When the output of a device spans such a wide range, temperature can be used to *tune* the wavelength.

Power delivered by these gems is anything but impressive, usually in the less than one to a few milliwatt range. Hence, many applications require arrays of lead salt diode lasers to generate enough illumination.

Perhaps the most significant drawback for these devices is the cryogenic temperatures needed for proper operation. In a laboratory setting, this is merely an inconvenience. But when it comes to practical applications in the real world, this surely presents problems. The notion of having to supply liquid nitrogen or helium to keep the laser cool doesn't thrill many industrial and/or design engineers.

AND, SOME OTHER WEIRD TYPES

In this section, I want to talk about two rather unorthodox concepts when applied to laser science (quantum electronics). The first is the *free-electron* laser and the second is the *x-ray* laser. Both are still experimental but have the potential to produce extremely high outputs.

FREE-ELECTRON LASERS

The free-electron laser depends upon the precept that high-energy electrons can be made to emit light if their path is altered by a strong magnetic field. This process involves shooting electrons down a column of varying magnetic fields which causes the electrons to oscillate back and forth.

This oscillation results in the formation of a light beam, similar to spontaneous emission, in the far to very-far infrared region. The mechanism is so efficient—20, 30, 40 percent—that beams of remarkably high power are possible. At least, in theory.

As seen in *Figure 11-6*, the three essential components of a free-electron laser are the *electron accelerator*, the *magnetic field* and the *optical resonance cavity*. At the risk of stating the obvious, each section contributes to the overall success of the laser. But, with these systems, that assertion carries far more significance.

Looking individually at these three sections, the *accelerator* can be one of a number of different approaches. These include linear type (electron guns), a cyclic unit (similar to a cyclotron) called a microtron, or a radio frequency based strategy that uses radio waves to propel electrons.

And there are even other tactics being explored in the lab. As with virtually all of the laser world, a great deal of research is aimed at improving the existing technology.

Once the electron beam has been produced, the next step is to subject it to a *magnetic field*. This field is not just any ordinary field. It has to be strong and result in an oscillating or wiggling effect that provides the needed alteration, or *bending*. This bending is what propagates the laser beam.

The field can be the product of either permanent or electromagnets and the size of the laser determines how many are required. The poles of the magnetic field are alternated to furnish regularly spaced control, and this system is sometimes dubbed a *wiggler* or *undulator*.

The last critical aspect of this laser is the resonance cavity that, like most lasers, provides amplification of the excited photons, yielding a population inversion. This consists of one partially reflective mirror and one fully reflective mirror, which should sound quite familiar by now.

When everything is put in its right place, a high-power source of collimated infrared laser light is born—one that allows for a great deal of modification and versatility, as well as a wide range of physical sizes.

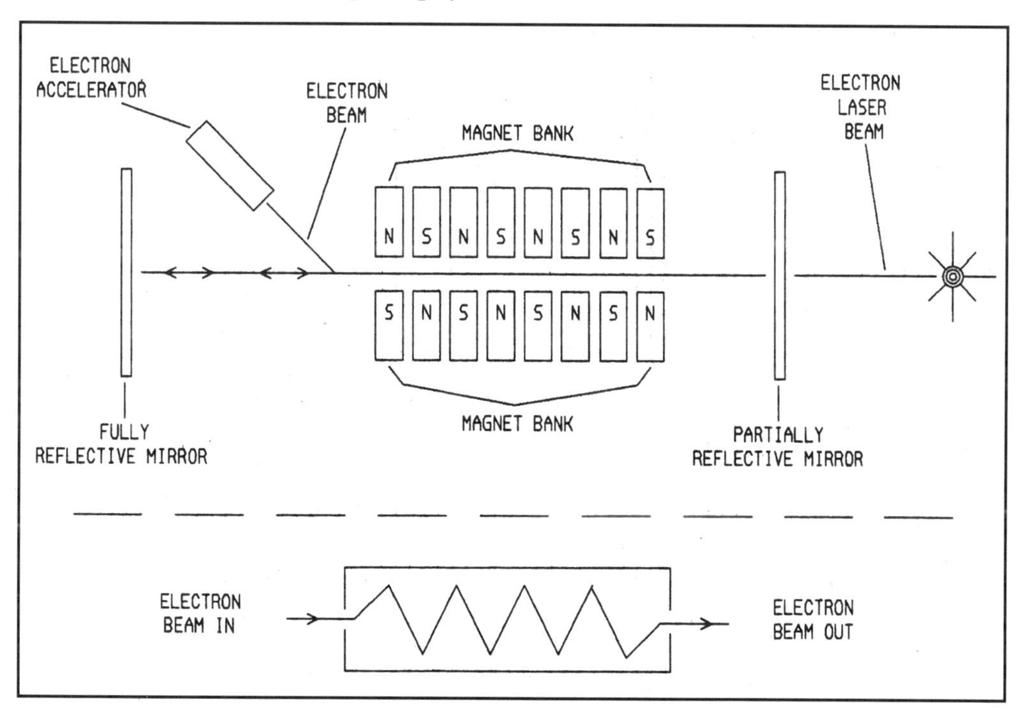

Figure 11-6. Free-electron laser. This diagram shows the "wiggle" effect created by the magnetic field.

Free-electron lasers are divided into two basic groups; the *Compton* devices and the *Raman* units. The principal differences between the two schemes concern the energy level and current. Raman lasers bear low electron energy and high current, while Compton devices display the opposite (low current/high energy).

Application of these systems is still in the infancy stage, but researchers continue to hunt for new purposes and better lasers. However, to date, spectroscopy, materials research and medical treatment lead the pack regarding device utilization. Additionally, the high power outputs attracted the attention of weapons developers such as the SDI program of the mid-80s.

Today, the free-electron laser is both a reality and a promising product of the future. Many of the factors discussed here have made the device appealing to the scientific community and commercial interests. However, a major drawback is expense. At this writing, the smallest of these units comes with a price tag in the hundreds of thousands of dollars. Hopefully, further development will bring that figure down.

X-RAY LASERS

Our last laser is the stuff Buck Rogers is made of. While producing a coherent, collimated beam of x-rays, these devices are a little far afield regarding laser science. They are also extensive projects with prohibitive costs unless you are a very large corporation or the federal government.

The x-ray laser functions on a principle that is somewhat opposite that of standard devices. Recall from Chapter 1 that photons are the product of electrons jumping from an inner orbit to an outer orbit then back again. In this process, the electrons absorb energy for the initial jump, then give it off as photons when they return to their normal position.

For x-rays, their production mandates an initial jump from an outer to an inner orbit. This requires a tremendous amount of energy, which is later expelled as x-rays. Hence, these lasers are capable of incredible output levels.

It was this factor, combined with the potentially small size of these systems, that again garnered notice by the weapons folks. Research was started in the early

1970s by the military, but due to the highly classified nature of that arena, little detail has emerged regarding x-ray lasers.

To date, there are two basic approaches to producing an x-ray beam. The first employs an inordinately high-powered external laser, while the second makes use of the energy spawned from a nuclear explosion. Both methods have been demonstrated by the Lawrence Livermore National Laboratory.

Using their Nova laser, the Livermore Lab was able to vaporize a thin strip of metal foil. This resulted in a plasma condition that produced a population inversion. Due to the massive energy levels involved, resonance mirrors were not required.

Wavelengths in the 4 to 32 nanometer range were displayed, but the huge Nova laser isn't exactly a pragmatic pump for a majority of x-ray laser experiments. Thus, as much additional research as economics will allow goes into developing smaller, more efficient pumping sources.

The second method is a little more exotic. Existing as the foundation of President Reagan's Strategic Defense Initiative (SDI), the *bomb-driven* x-ray laser utilizes the powerful x-rays from a nuclear explosion to do the metal strip/rod vaporization duties. As with the laboratory version, a plasma is created which leads to the population inversion.

The Lawrence Livermore National Laboratory was reported to have exhibited a working *bomb* laser in 1981, and later, with some prodding, confirmed that report. It emitted a wavelength of 1.4 nanometers, which is the shortest laser wavelength ever demonstrated.

Due to the classified status of the project, few exact details are available about this experiment, and/or its consequences. However, it is speculated that at least one such device does exist and is part of a defense program. Whether or not that is true could only be verified by the government.

If the technology has produced what it's supposed to have produced, these lasers have to be some of the most powerful, if not THE most powerful, ever built. Consequently, if the political hassles can be conquered, they should have a bright future.

CONCLUSION

And, thus ends the saga of other types of lasers. Bear in mind that I have really only scratched the surface here. There are additional designs out there, of less significance, and new ones coming along all the time.

But I do hope the information presented here has been informative, or at least entertaining, for you. Like so many concepts that start out as a mere idea, lasers have matured into a vast industry of both pragmatic and benevolent products.

I have often commented that the laser should be considered one of the top three inventions of the twentieth century, and I'll stick by that. Few discoveries have facilitated our lives in such a commanding and availing fashion.

On that note, let's move on to a quick and dirty history of the laser. By many standards, its chronicle is short, but that has not stopped it from impacting this world we live in.

CHAPTER 12

A SHORT HISTORY OF THE LASER AND ITS APPLICATION

As I mentioned in Chapter 1, lasers are a marriage of science understanding photon production, and the ever-fertile mind of Albert Einstein. However, Einstein did not live to see the first working laser, nor did he contribute to any of the early research (mostly masers) beyond his theory of stimulated emission.

The first laser(s) were the result of applying maser (Microwave Amplification by Stimulated Emission of Radiation) exploration to light wavelengths. And all of this was unquestionably enhanced by some additional fertile minds.

So, in this chapter we will look at the development of the laser and the players involved. It's an interesting story that I think you will enjoy! At least, I enjoyed writing it!

THE EARLY YEARS

Following Einstein's 1916 theory, over ten years passed before the first evidence of stimulated emission was observed (Landenburg, 1928). But that report did little to encourage substantial interest in the subject. For the next two decades, stimulated emission would remain in the laboratory.

However, the 1950s suddenly brought renewed attention to the phenomena in both the United States and the Soviet Union. This was due to maser research being done by Charles H. Townes (Columbia University), as well as exploration by Aleksandr M. Prokhorov and Nikolai Basov (Lebedev Physics Institute of Moscow).

Townes, with the help of Herbert Zeiger and James P. Gordon, is generally credited with the invention of the maser. He first conceived the idea in 1951 and the three had a working maser by 1953. The Soviet team completed the calculations for maser action in 1954, and Townes, Prokhorov and Basov shared the 1964 physics Nobel Prize for their work.

THE COMING OF THE LASER

With masers a reality, Townes and Bell Labs researcher Arthur L. Schawlow began looking into the possibility of applying stimulated emission to shorter wavelengths, such a light. They knew the requirements would be considerably different from microwaves and following extensive research, first patented the findings, then published them in 1958. One concept for lasers had thus been born.

However, they were not the only physicists concentrating on this task. Among the others was a Columbia graduate student by the name of Gordon Gould. By 1957, Gould had postulated many laser concepts still in use today, but failed to publish his work promptly. Rumor has it, Gordon considered himself an inventor as opposed to a scientist, and wanted to patent his ideas before letting the world in on his secrets. This attitude would end up bearing a cost in terms of prestige.

Due to the Townes/Schawlow patents, Gould faced many years of legal wrangling over patents for such things as Brewster windows, flash lamp excitation of solid-state lasers and electrical pumping of gas systems. Eventually, he did receive his patents in the late 1970s and 1980s, and most manufacturers to this day are licensed to employ his ideas in their laser products. But credit for the laser would go to others.

In a twist of fate, Gould actually benefited from the patent delay because it allowed the industry to catch up—catch up in terms of production and sales which heavily assisted his royalty payments. On a last Gordon Gould note, he IS given credit for coining the term *laser*.

Anyway, most of the early experimenters, including Townes and Gould, were convinced the best medium for laser action was *gas*. Gases would, of course, take a very prominent place in the cosmos of laser types, but a physicist working for the Malibu, California division of Hughes Research Laboratories thought differently.

His name was Theodore H. Maiman, and his preference was synthetic ruby. Maiman silvered both ends of a ruby rod (one end partially reflective) and pumped the device with a strong pulse of light from a flash tube. The result, in mid 1960, was the world's first working laser. Each time the lamp flashed, the ruby rod emitted a pulse of dark red light.

While others argued over the perfect laser medium, Maiman sedately produced his device and claimed the title *father of the laser*...well, at least in an informal way. Ironically, Maiman disobeyed company orders, as he had been told to cease work on his laser project. But in the end, the Hughes "brass" were all too happy to share in the success of the system, once operational.

The next several years would prove to be a furious race to invent and produce new types of lasers, and the scientists involved didn't disappoint anyone. In fact, most of the common lasers in use today were first introduced, in one form or another, during this period.

By the end of 1960 a calcium fluoride laser, doped with trivalent uranium, was developed by Peter P. Sorokin and M. J. Stevenson of IBM. Ali Javan, W. R. Bennett Jr. and Donald R. Herriott, all of Bell Laboratories, brought us the first helium-neon laser in 1961. This one operated in the infrared region, but the familiar red beam was soon to follow.

1961 also saw the first neodymium solid-state devices, in a calcium tungstate medium, from K. Nassau and L.F. Johnson, while Elias Snitzer developed the first neodymium glass laser. However, the YAG (yttrium aluminum garnet) systems, perfected by H. M. Marcos, L. G. Uitert and J. E. Geusic, would be two years in coming.

1962 brought the first semiconductor lasers. These were the work of three different research organizations; MIT's Lincoln Laboratories, IBM's Watson Research Center and General Electric Research Laboratories. All three produced gallium arsenide devices that had to be operated at cryogenic temperatures. Usually, liquid nitrogen was used as the cooling agent.

And, so it went! One device emerging on the heels of another as the world watched the laser field *explode* before its very eyes.

WHAT TO DO WITH THAT PRETTY RED DOT

By the mid 1960s, lasers were becoming almost commonplace and research into new and different methods created a "blurrrr" regarding chronology. But just as important, lasers were beginning to be put to practical use. As scientists began to comprehend the properties of coherent light, its value became even more apparent.

All of this inspired another race, between a number of manufacturing interests, to find applications for this newfound wonder. While fueled by profits, these efforts helped establish the laser as one of the most important inventions of the twentieth century. It is unlikely lasers would have reached the success and prominence they enjoy today if not for those commercial energies.

One of the first areas to benefit from the laser was the medical profession. With the advent of higher-power systems, many fields of medicine found a new friend, especially in the domain of surgery. A characteristic of human tissue, learned early on, is that it absorbs and diffuses infrared light. This allows for a carbon dioxide laser, capable of cutting a steel plate in half, to be used for delicate surgical techniques.

Other medical applications involved more precise internal procedures and very controlled heating. The dental profession found this second aspect abundantly handy. For example, special plastic materials can be used for fillings, in place of the conventional mercury/silver amalgam, and cured with the heat from a laser.

So, the medical world definitely profited by the introduction of this device and continues to do so. However, that was by no means the only area to gain. In fact, you can see lasers at work all around you. Sometimes, it just takes a close look to recognize their presence.

One of the more overt arenas is the entertainment industry. Boy did the laser find a home here! Laser light shows are used as part of performances, or performances all their own. A combination of various laser types (and wavelengths) allows for a variety of spectacular visual effects. Add computer control, and you have programmed effects, as well as full animation.

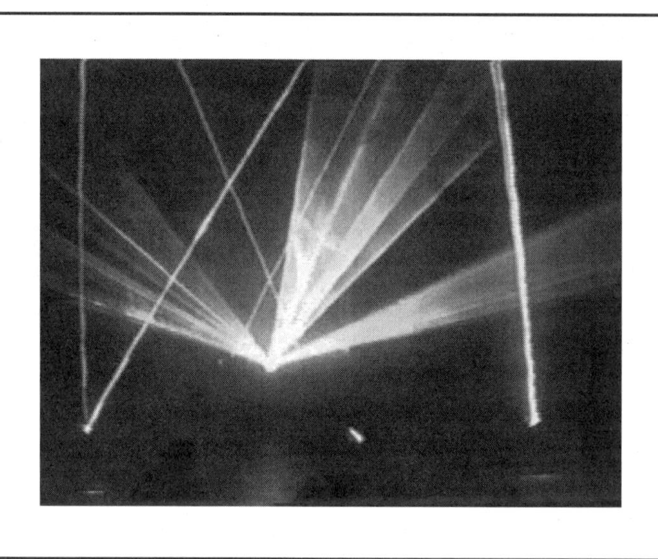

Figure 12-1. A laser light show, compliments of the copper laser. This laser recently accompanied the Pink Floyd world tour. (Photo courtesy of Oxford Lasers)

Additionally, Hollywood makes use of lasers in a more behind-the-scenes fashion. When building sets for both television and motion pictures, they are extremely handy leveling tools. And, they have been employed as very discrete (except to the actors) stage marks that keep everybody in the right place.

Another *overt* area where lasers shine (excuse me), is publishing/printing. The collimation experienced with these devices provides extremely good resolution. Resolution that often rivals an old fashion printing press. And, let's not forget the laser-based scanners. Again, the quality of product is so high that it takes a real close gander to tell the difference between a traditional photographic print and a laser scan.

While we're on the subject of digital electronics, the compact disk (CD), compact disk read-only memory (CD-ROM) and optical data reader/writer would all be lost without lasers. In theory, a beam of light can be focused down to its wavelength, thus allowing for a very tiny spot of light. Those tiny spots further allow for very small areas to be illuminated, and that is the secret to the huge amount of data found on CDs/CD-ROMs.

Figure 12-2. CD players

The spots, or *pits*, burned into the aluminum foil base that represent the data on a CD are so small it takes an electron microscope to see them. They are lined up in a track, much like the groove on a phonograph record, and if you were able to stretch that track out, it would extend to around four and a half miles.

Of course, focusing the beam to its wavelength is theory, and theory doesn't always pan out in practical application. More to the point, it rarely pans out in practical application. But, the natural collimation of laser light gets us closer to that theoretical goal than any other form of illumination.

Hence, using light amplification by stimulated emission of radiation (just wanted to see if you were still awake), the beam can literally be focused down to a microscopic pinpoint. That furnishes the process for placing a tremendous amount of information on such a small round piece of plastic.

Next! I touched on this domain briefly, but let me expand a tad. When it comes to drawing a straight line, you will have to look far and wide for anything that does a better job than a laser beam. And there lies a value for the construction industry.

Not only can they be used to level everything from a foundation to a linear accelerator, lasers also help square up corners, keep walls straight and provide straight edges for cutting large sections of wood, wallboard, paneling, et cetera.

All of this makes for better construction—or at least it is supposed to. There are times, however, when I wonder if the contractor has lost his laser level. Anyway, when properly employed, the laser can be a real friend to the construction folks.

Along these same lines, the high-power carbon dioxide units are routinely employed in metal and plastic fabrication. They provide an extremely precise cut and are wizards at welding and edge sealing. Hence, the automotive, heavy construction and metal/plastic manufacturing industries have benefited immensely from this technology.

And, while discussing commercial applications, let's not forget marketing. Not only are lasers used to attract attention, but they also help keep a store supplied (universal product code (UPC) scanning). Additionally, they help get you out of the place. Who has not encountered a UPC bar code reader at the checkout counter.

Regarding the scientific community, I'm not even going to begin to try to cover every aspect aided by lasers in this area. From chemistry to physics to optics to geology to just about any field you can imagine, lasers have become an important lab and field tool.

Furthermore, these miraculous "flashlights" assist both researchers and engineers alike. Researchers can use them for study purposes, while engineers find lasers very handy at the production end. In either case, the final result is usually better off for the effort. Ooops, there is more of that theory.

With the arrival of optical fiber (or should I say *reliable* optical fiber), the telecommunications industry was quick to recognize the value of collimated light. Today, optical links are a common part of both telephone and television systems, and there is no foreseeable end to this practice.

Using light to transmit information has a number of very distinct advantages. First of all, as opposed to electrical lines, the optical lines are extremely quiet. Much of that all-too-familiar static, and other assorted noises, simply isn't found on fiber optics.

This part is very good news to anyone relying on data transfer over the telephone. Quiet lines are highly conducive to faster, more dependable communications. And that translates into greater efficiency, productivity, simplicity and PROFITS!

Alright! Let's talk about semiconductor lasers. While they date back to the early '60s, it has only been in the last ten years or so that *diode* lasers have become viable entities outside the laboratory. The initial efforts at these devices resulted in operational temperatures in the minus 273 degree range.

Additionally, they required high currents and/or were extremely inefficient, all characteristics that didn't make them desirable to commercial interests. However, that has all changed. Most of today's laser diodes operate at room temperature, with 3 to 6 volts and current levels of 25 to 50 milliamps.

Due to this progress, small semiconductor lasers have found their way into a variety of applications and products. The laser pointer for instance! Ah yes, where would we be without our laser pointers?

At this time, the red wavelength diodes, in the 670 to 635 nanometer range, are giving the classic helium-neon tubes a real run for their money. And, new high-power infrared devices, with outputs up to one watt, will certainly find a number of niches in both the scientific and commercial communities.

CONCLUSION

And, that's the tale. At least part of it. As with all technology, the story still has many chapters to be written. But, I hope I have given you the highlights of how we arrived at this fascinating world of coherent and collimated light.

Research into newer and more efficient lasers is going on as we speak, and that will, of course, lead to additional product and applications. Thus, there are few foreseeable boundaries concerning the ultimate roads the laser industry will travel.

In the end, however, advancements in this area will depend largely on requirements and technological expertise. Take the case of cellular telephones. Scientific achievement had to catch up with the concept. But, the old adage, "necessity is the mother of invention" has always served us well. And, I expect laser progress will follow suit.

Don't be surprised at what is just around the corner! Regarding this field, I suspect there are numerous revelations awaiting our discovery. Many, I'm sure, will simply astound us!

CONCLUSION

Well gang, we've come to the end of the trail. But, it has been a delightful journey! One that I always enjoy taking! The road is smooth, not many potholes, and the destination is forever a pleasant one. Don't you just love how I work in all these "travel" allusions?

Anyway, I do hope you appreciated the expedition (there I go again). OK, enough....enough! But, I do have to have my fun. However, I've gotten it out of my system. I promise.

Light amplification by stimulated emission of radiation (LASER) has had a brief but illustrious history. As we all know, it emerged as a short pulse of deep red light from the end of a synthetic ruby rod and has gone on to become one of the most significant inventions of the twentieth century.

Throughout this chronicle, the laser has undergone many a transformation, rebirth and metamorphosis. It has branched off in many different directions and used a multitude of diverse materials to evolve into what we have today. And, all of this has been to our benefit.

At this time in our history, lasers are capable of performing some downright spectacular tasks. Carbon dioxide devices are used for delicate eye surgery but can also cut a thick steel plate in half. Helium-neon tubes run the gamut from reading universal product codes (UPC) at the grocery store to providing entertainment as part of vibrant and sensational light shows.

Other applications include three-dimensional photographs (holograms), a vast expanse of scientific research, weapons, communications, aligning tunnels, foundations, walls, etc. when employed in the construction industry, hardening of filling in dentistry and emphasizing your point during a presentation (laser pointers). And, this is just the beginning. I could go on and on, but I would probably run out of space before I ran out of uses for this marvelous technology.

Throughout the text, we have talked about some of these applications, and I hope this discourse has lived up to your expectations. As I stated at the outset, I have an ongoing fascination with the laser that dates back to its 1960 inception. And, I can't think of too many topics I prefer to talk about. Hence, this has been a very gratifying experience for me. I mean, you get me going on this subject, and I'll likely yak your ear off. But, this is truly something worth talking about, so I hope this book inspires you, too, to do some vocalizing.

Well, it's time to check out. I thank you for making the *journey* with me. I know...I know! I promised not to allude to travel anymore. I guess I just can't be trusted. Oh well, go have some fun with lasers. They can and will generate a lot of it if you give 'em a chance.

ACKNOWLEDGMENTS

I would like to thank the following for their contributions in producing this text. It was greatly appreciated!

Thanks!

Carl

Compton's Interactive Encyclopedia, 1994 Edition

Gernsback Publications

How to Make Holograms
By Don McNair
Tab Books, Inc.
1983

The Laser Guidebook
By Jeff Hecht
McGraw-Hill, Inc.
1986

SOURCE LIST

This list represents many of the companies I have used in the past. They carry parts, supplies and other items needed to build the projects in this book. These, of course, are not all the available distributors, and you may well know of others.

For each source, I included their address, as many phone numbers and other means of communicating with them as I could find, and a brief description of what they carry. Hopefully these will be helpful concerning the devices in this book.

1) ALL ELECTRONICS CORPORATION

Post Office Box 567

Van Nuys, CA 91408-0567

Phone: (800) 826-5432

Good line of surplus gear as well as useful parts.

2) ALTRONICS

2300 Zanker Road

San Jose, CA 95131

Phone: (408) 943-9773

Fax: (408) 943-9776

Laser associated stuff, enclosures, switches and other parts.

3) AMERICAN SCIENCE & SURPLUS

3605 Howard Street

Skokie, IL 60076

Phone: (847) 982-0870

Fax: (800) 934-0722

Tremendous source of surplus parts, HeNe lasers, optics, hardware, power supplies, etc.

4) B.G. MICRO

Post Office Box 280298

Dallas, TX 75228

Orders: (800) 276-2206

Tech: (972) 271-9834

Fax: (972) 271-2462

Internet: http://www.bgmicro.com/

E-Mail: bgmicro@bgmicro.com

Good source for surplus equipment and parts. Highly customer service oriented with extremely fast delivery.

5) THE C-THRU RULER COMPANY

Post Office Box 356

Bloomfield, CT 06002

Orders/Info: (860) 243-0303

This is the source for the dry transfer lettering if you can not find it locally. Excellent lettering.

6) CIRCUIT SPECIALISTS, INC.

Post Office Box 3047

Scottsdale, AZ 85271-3047

Orders: (800) 528-1417

Info: (602) 464-2485

Fax: (602) 464-5824

Internet: http://www.cir.com

Good line of ICs, components, hardware and other supplies.

7) DC ELECTRONICS

Post Office Box 3203

Scottsdale, AZ 85271-3203

Phone: (602) 945-7736

Orders: (800) 467-7736 or (800) 423-0070

Fax: (602) 994-1707

Good line of specialized ICs and transistors.

8) DEBCO ELECTRONICS

4025 Edwards Road

Cincinnati, OH 45209

Orders: (800) 423-4499

Info: (513) 531-4499

Fax: (513) 531-4455

Carries Rainbow kits as well as many parts.

9) DIGI-KEY CORPORATION

701 Brooks Avenue South

Thief River Falls, MN 56701-0677

Orders: (800) 344-4539

Info: (218) 681-6674

Fax: (218) 681-3380

Internet: http://www.digikey.com

Large catalog of resent ICs, parts, laser diodes, coils, switches, discrete components, etc.

10) ELECTRONIC GOLDMINE

Post Office Box 5408

Scottsdale, AZ 85261

Phone: (602) 451-7454

Fax: (602) 661-8259

Internet: http://www.goldmine-elec.com

Surplus house that carries parts, hardware, cases, etc.

11) HOSFELT ELECTRONICS, INC

2700 Sunset Boulevard

Steubenville, OH 43592

Order: (800) 524-6464, (888) 264-6464

Phone: (740) 264-6464

FAX: (800) 524-5414

Excellent line of laser diodes, diode modules and pointers as well as other electronic components.

12) JAMECO ELECTRONIC COMPONENTS/COMPUTER PRODUCTS

1355 Shoreway Road

Belmont, CA 94002-4100

Phone: (415) 592-8097

Order: (800) 831-4242

Fax: (800) 237-6948

Internet: http://www.jameco.com

Full service distributor for parts, cases, hardware and most everything related to this text. Good selection of diode lasers, collimators and drivers.

13) JDR MICRODEVICES

1850 South 10th Street

San Jose, CA 95112-4108

Phone: (408) 494-1400

Order: (800) 538-5000

Fax: (408) 494-1420

Internet: http://www.jdr.com

Full line of parts, cases, laser diodes, hardware and other useful items.

14) MARLIN P. JONES & ASSOCIATES, INC.

Post Office Box 12685

Lake Park, FL 33403-0685

Phone: (800) 652-6733

Fax: (800) 432-9937

Internet: www.mpja.com

E-Mail: mpja@mpja.com

Extensive inventory of parts, hardware, transformers, etc. Carries high-voltage power supply kit for Chapter 3.

15) MCM ELECTRONICS

650 Congress Park Drive

Centerville, OH 45459-4072

Phone: (800) 543-4330

Good line of useful parts, hardware, etc. for projects.

16) MENDELSON ELECTRONICS COMPANY, INC. (MECI)

340 East First Street

Dayton, OH 45402

Orders: (800) 344-4465

Fax: (800) 344-6324

Carries parts, hardware and other surplus items.

17) MEREDITH INSTRUMENTS, INC.

5420 West Camelback Road, #4

Post Office Box 1724

Glendale, AZ 85301

Order: (800) 722-0392

Info: (602) 934-9387

FAX: (602) 934-9482

Internet: http//www.mi-laser.com

Specialty source for lasers of many different types, sizes, power ratings, etc. and power supplies, optics and other laser associated items.

18) MIDWEST LASER PRODUCTS

Post Office Box 262

Frankfort, IL 60423

Phone: (815) 464-0085

FAX: (815) 464-0767

Internet: www.midwest-laser.com

Another well stocked laser specialty outlet. Good prices on HeNe, laser diodes, argon lasers and other related stuff.

19) MOUSER ELECTRONICS

National Circulation Center

2401 Highway 287 North

Mansfield, TX 76063-4827

Phone: (800) 346-6873

Internet: http://www.mouser.com

E-Mail: sales@mouser.com

Extensive line of parts, chemicals, PC board materials.

20) MWK INDUSTRIES

455 West La Cadena #14

Riverside, CA 92501

Phone: (909) 784-4888

FAX: (909) 784-4890

E-Mail: mwk@worldnet.att.net, mkenny1989@aol.com

Internet: www.mwkindustries.comm

Very complete specialty source for lasers and laser related items. They carry many types of lasers, optics, power supplies, light shows and more.

21) RADIO SHACK

Local Stores

Tandy Corporation

Ft. Worth, TX 76102

Handles a limited line of parts and hardware. Good source of switches and cases.

22) TECH AMERICA

Post Office Box 1981

Fort Worth, TX 76101-1981

Orders: (800) 877-0072

Fax: (800) 813-0087

Tech Hot Line: (800) 876-5292

Cust Serv: (800) 613-7080

A tremendous catalog full of parts, hardware, enclosures and just about everything you will need.

23) TIMELINE INC.

2539 West 237th Street

Building F

Torrance, CA 90505

Orders: (800) 872-8878

Tech.info: (310) 784-5488

Fax: (310) 784-7590

Good line of surplus items, including laser diodes, chips, LCD displays and other useful stuff.

GLOSSARY

AAA

ADC	Analog to digital converter. A device that changes digital signals to the analog state.
Alternating Current	(AC) A voltage source that repeatedly changes polarity. 60 hertz household line is an example.
Amperage	Named after French physicist Andre Marie Ampere. Also called "current," this is the "force" behind a voltage. May be denoted as Amps, Milli-, or Microamps.
Amplifier	A circuit that increases the gain of an input signal such as audio. Also, with lasers, it can be a glass element for solid-state systems.
Analog	Any signal that contains continuous information as opposed to "0s" and "1s". Radio, audio and TV are good examples of this form of signal.
Analog Meter	The older "moving pointer" style meter that utilizes an inductive movement.
Anode	The positive (+) side of a polarity-sensitive component such as a diode.
Audio	Referring to anything in the human ear sound range of frequencies.
Argon Laser	A gaseous laser that uses argon gas as the medium. These are some of the most powerful lasers.

BBB

Ballast Resistor	A resistor used to balance the anode input of a helium-neon tube.
Basov, Nikolia	A Russian scientist from the Lebedev Physics Institute (Moscow) involved in early maser research.
Beam Divergence	The amount a laser beam spreads over a specified distance.
Beam Splitter	An optical device that splits a light beam into two separate beams.
Bennett, W.R. Jr.	One of a team of three Bell scientists involved in developing the first helium-neon laser.
Bleach	No, not Clorox. One of the chemicals used to process holograms.
Bore	The core of a helium-neon, and some other lasers that largely determines the diameter of the output beam.
Breadboard	A systematic system for laying out prototype circuits. The solderless breadboard is a prime example.
Brewster Window	A special "angled" optical lens that is used at the output aperture of several different laser types.

CCC

C.D.R.H.	Center for Device and Radiological Health, a government agency that sets the standards for laser safety.
Calibrate	Any process that aligns an operation or measurement with the expected value. Adjusting the reference potentiometer on a digital voltmeter, for example.

Capacitor	A common electronic component that has the ability to store a charge for a short time. Also used to prevent direct current (DC) flow.
Carbon Dioxide Laser	A group of high power lasers that use carbon dioxide gas and are used for, among other things, metal fabrication and medical surgery.
Cathode	The negative side of a polarity-sensitive component such as a diode.
Cavity	A term for the resonance structure used with ruby rod and other solid-state lasers.
Chip	Slang for "integrated circuit".
Coherent Light	One property of laser light that keeps the beam together. There are two different forms, spatial and temporal.
Coil	An electronic component comprised of wired wound around a center core. This device produces a magnetic field and is often called an electromagnet.
Cold Joint	A poorly executed solder joint that is characterized by a dull "flaky" appearance, and poor conductivity.
Collector	The side of a transistor junction that receives the current flow.
Collimation	The property of laser light that prevents it from spreading like other forms of light.
Collimator	A device used with lasers to assist or improve collimation. Usually a lens assembly.
Communications	The process of conveying information from one point to another.

Complimentary Metal Oxide Semiconductor	(CMOS) A type of integrated circuit (IC) architecture that provides low power consumption over a wide range of voltages. Frequency response is dependent on operational voltage with a limit of about 10 megahertz.
Controller	Any device that regulates operation of another device or function.
Converter	A circuit that changes a signal from one electronic state to another. Analog to digital is a good example.
Concave Lens	Any lens where one or both sides curve inward as opposed to outward.
Continuous Beam	A laser beam that is constant as opposed to pulsed. Helium-neon lasers are a good example.
Convex Lens	Any lens where one or both sides curve outward as opposed to inward.
Copper Vapor Laser	A laser that uses copper vapor as its medium. The copper is vaporized with high temperature heating elements.
Coupling Capacitor	Employing a capacitor in between two stages to provide smooth interaction of the stages. Also, usually performs DC voltage blocking.
Crystal	An electronic component made from quartz that resonates when electricity is passed through it. Used primarily for frequency control.

DDD

DAC	(digital-to-analog converter) This device takes a digital string of "1"s and "0"s and changes it into an analog signal.
DC Block	Any method that prevents direct current (DC) flow. Capacitors are the most commonly used device for this purpose.

Decode	In this context, this is the receiver of a light beam communications link interpreting the signal.
Developer	The chemical used in hologram processing that produces the image.
Digital	Referring to forcing the value of data to either a zero (0) or a one (1), as opposed to infinitely many values between zero and one. This process allows microprocessors to evaluate and process said data.
Digital Multimeter	(DMM) An electronic testing device that incorporates several different measurement tools. Common DMMs can measure voltage, amperage, resistance, with some adding frequency and capacitance to the equation.
Diode	An electronic component that rectifies alternating current (AC) to direct current (DC). Also used for radio signal detection.
Diode Laser	A semiconductor device similar to a light emitting diode that produces true laser light. Used in "laser pointers".
Direct Current	(DC) A voltage source with a fixed polarity. Power supplies and batteries are two primary examples.
Display	A device used to visually provide information. LED and LCD are two technologies commonly used for this purpose.
Doping	A process of introducing impurities into glass or YAG to act as the laser medium. Also applies to laser diodes.
Double Hetero-Junction	The latest of the processes used to make semiconductor lasers.
DPDT	(Double Pole—Double Throw) Designates a switch that can "throw" two signals in one of two directions.

DPST	(Double Pole—Single Throw) Designates a switch that "throws" two signals in only one direction. In effect, a double "on-off" switch.
Duplex	A term used to describe 2-way communications where both ends can hear and talk simultaneously.
Duty Cycle	This term is used to characterize the ratio between the "crests" and "troughs" of an output signal or cycle.
Dye Laser	A laser type that utilizes fluorescent dyes as their medium. Allows the laser to be tuned to many different wavelengths.

EEE

Einstein, Albert	One of the foremost theoretical physicists of all time. Lasers are founded on his theory of "stimulated emission of radiation."
Electromagnet	Device that generates a magnetic field by passing electricity through a coil of wire. As opposed to a permanent magnet.
Electret Microphone	A type of microphone that uses a capacitor as the "pick-up" element. By detecting the change in capacitance, due to vibration, the mic is capable of perceiving sound with high sensitivity.
Electron	One of the particles that make up atoms. Electrons orbit the center or nucleus.
Emitter	The side of a transistor junction that supplies current flow.
Encode	In this context, the laser modulator that transmits the light beam.
Excited State	An atomic state that exists when an electron has jumped to an outer orbit. This process mandates the electron absorbing energy.

FFF

Farad	Named after Michael Faraday, this is the basic unit of capacitance.
Field Effect Transistor	(FET) Specialized semiconductor characterized by very high input resistance.
Filter	Plates or circles of glass or plastic that restrict the passage of certain wavelengths and pass others.
Fixer	The chemical in hologram processing that removes excess "silver based" emulsion, making the hologram permanent.
Flash Tube	A sealed glass tube, usually filled with xenon gas, that has electrodes at each end. When a high voltage is applied to the electrodes, the tube emits a strong flash of white light.
Fluorescent Dye	A series of organic dyes that fluoresce in the presence of ultraviolet light. However, for our purposes, they are more important as the medium of many liquid lasers.
Fog	Those low clouds that role into San Francisco. Well, it is that, but it is also interference caused by unwanted exposure from stray light. Interestingly enough, in holograms it almost looks like that stuff in San Francisco.
Free-Air Communications	Light communications done by sending the light through an open space of air instead of over a fiber optic link.
Free-Electron Laser	Type of laser in which strong magnetic fields are used to "wiggle" a stream of electrons back and forth which causes them to emit coherent light.
Frequency	Measured in hertz, it represents the number of "cycles" that occur in one second. Can be kilohertz, megahertz, gigahertz and so forth.

Full Duplex — Refers to the ability of a communications system to both send and receive simultaneously. A standard "landline" telephone is an example.

GGG

Gallium-Arsenide Junction — Combination of the metal gallium and the metalloid arsenic, used by most laser diodes and light emitting diodes. Most junctions also add trace amounts of other elements.

Gas Out — A laser term that means a gaseous laser has a leak and is letting gas escape.

Gaseous — Refers to a class of lasers that use gas as the laser medium. Beyond that I will not go.

Geusic, J.E. — One of three scientists responsible for development of the YAG laser.

Glass Rod Laser — Any one of a number of lasers based on a glass rod. In many cases, the glass has trace amounts of neodymium and/or other elements as the medium.

Gordon, James P. — One of two scientists who helped Charles Townes develop the maser.

Gould, Gordon — Columbia University grad student who postulated many of the concepts still used by today's lasers. He also holds many patents that affect laser manufacturing.

Ground — A common "rail" or "leg" in a circuit, usually negative in polarity, that provides collective feedback for the circuit. Often, a substantial number of circuit components will have at least one lead connected to ground.

HHH

Haloing	An optical effect of laser beams in which other wavelengths present are displayed as rings around the main beam.
Hard Seal Tubes	A method of sealing gaseous laser tubes by fusing the mirror holders and electrodes directly to the molten glass envelope. As opposed to gluing the on. Makes for a far more secure seam that produces longer life.
Helium-Neon Laser	Easily the most common gaseous laser, it uses a helium gas and neon gas combination as the medium. It is also one of the most common hobby lasers.
Henry	Named after American Joseph Henry, it is the basic unit of inductance.
Herrott, Donald R.	One of three Bell Lab scientists responsible for development of the helium-neon laser.
Hertz	Honoring German electrical pioneer Heinrich Hertz, it is the basic unit for frequency, or cycles per second.
High Dielectric	A term used to describe hookup wire that can handle high voltages.
High Voltage Supply	A power supply that produces the high voltages and currents necessary for gas lasers and xenon flash tubes.
Hologram	A three-dimensional photograph that uses a laser for the illumination.
Holography	The science of using lasers to produce true three-dimensional photographs.
Homojunction	The original "one piece" architecture used for semiconductor, or diode lasers. Pretty much obsolete.

III

Impedance — Gauged in Ohms, this is the inductive resistance to the flow of alternating current (AC). Usually refers to speaker coils or antennas.

Inductor — Another name for a coil, or device in which magnetic fields are created by current flow.

Infrared — A spectrum of invisible light that extends off the "red" end of the visible light spectrum.

Input — The signal put into a circuit, or the place where the signal is put into said circuit.

Integrated Circuit — (IC) Any one of a huge number of active devices in which an entire circuit is etched onto/into a silicon wafer. The result usually allows complex equipment to be built with a minimum of components. It also greatly enhances the reliability of the finished device.

Interface — Any arrangement that allows one circuit to connect to another. Can be as simple as just a cable, or as complicated as an entire third circuit.

Interference Pattern — The interaction of two or more light beams that is the "heart and soul" of holography.

JJJ

Jack — Any one of numerous receptors that accepts a matching plug. For example, an earphone jack on a radio.

Javan, Ali — One of three Bell Lab scientists responsible for the development of the helium-neon laser.

Jerk — A person who fits into the categories of "selfish, inconsiderate, stupid, obnoxious or all the above." Sort of like that idiot that tried to crawl over your wall but you caught with your perimeter alarm and had arrested.

Johnson, L.F. — One of two scientists instrumental in developing the neodymium lasers.

KKK

Kicker Voltage — A term for the initial jolt of very high voltage necessary to get a helium-neon laser, and some others, going.

Krypton Laser — A gaseous laser that uses krypton as the laser medium. Produces a red beam.

LLL

Laser — (Light Amplification by Stimulated Emission of Radiation) An optical device that produces a coherent, collimated beam of light. Usually of primarily one color, this light can travel long distances without much beam spread or intensity loss.

Laser Beam — The thread-like ray of monochromatic light that emerges from a laser. Can be visible or invisible light.

Laser Diode — A specialized semiconductor diode that produces true laser light. These have become extremely popular as the light source for fiber optic communications. The light is easily modulated and bears all the properties of more standard laser illumination.

Laser Head — Refers to a laser tube that is contained in a tube or other structure. This is done for both tube protection and safety against shock.

Laser Medium	The material present, whatever it might be, that produces the "population inversion" that leads to the creation of a Laser beam.
Laser Tube	The glass envelope that contains the gas used as the laser medium. Usually also supports the electrodes, and sometimes the resonance mirrors.
Latching Relay	A type of mechanical relay that stays in position after the magnetic coil is deactivated. Another signal is used to release it.
Light Emitting Diode	(LED) A specialized diode who's junction glows when an electrical current is passed through it. The junctions are often a combination of arsenic and gallium, and can produce light of just about every color.
Line of Sight	Refers to the characteristic of light carrying only as far as can be seen.
Linear Regulator	One of the most common ICs used to regulate voltages. While not especially efficient, they are highly accurate and easy to use.
Liquid Laser	Any laser that uses a liquid as the lasing medium. The most common example is the dye laser that uses liquid fluorescent dyes.
Liquid Crystal Display	LCD for short, these are display arrays that use chemicals known as "liquid crystals." These chemicals have the property of changing color when their molecules are "twisted" by electricity and become visible in whatever pattern they are arranged.
Lissajous Pattern	A term used to describe light patterns produced by laser light shows. These can include randomly changing lines, dots and other designs.

Load	The device being controlled by a "controller" circuit. A relay, motor, LED, SCR as a few examples.

MMM

Maiman, Theodore	A Hughes Research Laboratory scientist that developed the first working laser. He did so in Malibu, California, in 1960, with a rod of synthetic ruby.
Marcos, H.M.	One of three scientists responsible for the YAG type lasers.
MASER	Acronym for "Microwave Amplification by Stimulated Emission of Radiation," it was the forerunner of the laser and still exists today.
Meniscus Lens	Lens assembly of two or more optical elements. Sometimes called a "compound lens."
Microphone	Any of many differently designed electronic sensors that detects audio sound.
Milliwatt	One thousandth of a watt, this unit is used to measure and rate the power of many laser types.
Mirror	Any device that reflects light. These can be full or partially reflective and flat or curved.
Modulation	The process of attaching information to any "carrier" style signal. This can be by wire, light or radio waves.
Monochromisity	The property of some light being predominately of one wavelength. An essential part of laser light, the 632.8 nanometer wavelength of some helium-neon lasers is a good example.

NNN

Nanometer	One billionth of a meter, this is a frequently used unit of measure for light wavelengths.
Nassau, K.	One of two scientists that developed the neodymium solid-state lasers.
Negative Lens	A lens characterized by one or both surfaces curving inward. Used to spread the laser beam in holographic setups.
Neodymium	A "rare earth" metallic element that functions well as the active medium for both glass and synthetic crystal lasers. Symbol (Nd).
Neon Lamp	An "ionization" lamp where electricity applied to electrodes in an envelope of neon gas causes the neon to glow.
Non-Inverting	Any device, chip, system that doesn't change the state of the input signal. Such as a "buffer" gate or one input of an operational amplifier.
Nitrogen Laser	A gaseous laser that uses nitrogen as the active medium. One of the simplest lasers to build.
Nova Laser	A huge neodymium glass laser located at the Lawrence Livermore National Laboratory. Considered by many to be the most powerful solid-state device to date, it has been used for experiments in areas such as scientific research and weapons.

OOO

Object Beam	In holography, this is the beam that lights the subject (object). In many "split-beam" arrangements, it can actually be more than one beam.

Ohm	Named for German mathematician George Simon Ohm, this is the basic unit of resistance.
Operational Amplifier	A series of amplifier integrated circuits that are characterized by inverting/non-inverting inputs, high gain and some times split power supply requirements.
Optics	Any of a family of devices that control, modify or otherwise affect light. Lenses, filters and prisms are three examples.
Orbital Jump	The process of an electron jumping from its natural orbit to another orbit within an atom. Energy is required for the jump and the electron will eventually return to norm, releasing the energy as a photon.
Oscillator	A device that produces a repeating cycle pattern at a specific frequency. The LM555 IC is an example of an oscillator circuit.
Output	The signal emitted by devices such as amplifiers, oscillators, power supplies, et cetera.

PPP

Partial Mirror	A mirror that has an incomplete reflective coating. These mirrors allow some light to pass through them and are used for resonance in lasers and as beam splitters.
Perf-Board	A circuit construction material that has a uniform pattern of holes for component installation. Due to standard lead spacing, the holes are often 0.1 inches apart. Some boards also have copper soldering pads on one side for easier parts placement.
Phototransistor	A type of transistor where light controls the current flow instead of an electric potential on the base.

Photoresistor	A type of resistor where light changes the value of the device.
Photon	The atomic particle that makes up both visible and invisible light.
Plasma	An electrically neutral ionized gas or the science of such. Often under high temperature.
Plug	The "male" half of any specific receptor system. The power plug on your computer for instance.
Point-To-Point Wiring	The process of connecting electronic components with individual wires as opposed to the copper "traces" of a printed-circuit board.
Polarity	With DC oriented electronics, this refers to the positive (+) and negative (-) poles of the power source or components involved.
Population Inversion	The point at which one more than half the atoms in a specified quantity are excited. This is a critical occurrence for the production laser light.
Positioning Table	The structure used to set up holograms. Usually characterized by heavy weight and extreme resistance to vibration. Many use sand for both weight and to position the various optical elements needed.
Positive	The plus (+) pole of a battery, DC supply, electrolytic capacitor or other polarity sensitive item.
Positive Lens	A lens with one or both sides curved outward. These lenses condense light as opposed to spreading it.
Potentiometer	A variable form of resistor that allows a change in resistance. Volume controls are a prime example.

Power Supply	Any device that provides the working potential for anything electric or electronic. Can be a battery, transformer based supply, generator and so forth.
Power Transistor	A specialized transistor designed to handle high voltages, current or both.
Printed-Circuit Board	(PCB) A fiber- or epoxy-based circuit board that has copper traces etched on one or both sides for conduction of the circuit's electrical power. Makes for very stable and sturdy circuit construction.
Prism	An optical device that bends, reverses and/or transposes light waves. Equilateral and pentaprisms are two examples. Some can break white light up into its spectral components.
Prokhorov, Aleksander M.	Russian scientist involved in early maser research.
Prototype	The first, or experimental version of a design or idea. With electronics, this often begins on a solderless style breadboard.
Pulse	A short burst of light usually from a solid-state laser.
Pulse Width Modulation	A form of modulation that utilizes duty cycle to control the signal. In other words, on/off time of the cycle dictates the operation of another component.
Pumping	Refers to applying energy to a laser medium to generate a population inversion. Electricity and powerful flashes of light are the two most common methods.

QQQ

Quantum Well	A micro-micro-microscopic depression style junction, only a few hundred atoms wide, that is responsible for the high efficiency of the newer laser diodes.

Quartz	A form of silicon dioxide (SiO2), it is the preferred choice of material for crystals and piezo elements.

RRR

R/C Tank (Tuning)	A method of tuning an oscillator in which a resistor (R) and capacitor (C) are employed in a parallel arrangement to set the frequency.
Receiver	Refers to the end of a light wave communications link that "hears" the transmitted light beam.
Rectifier Bridge	A full-wave rectifier circuit contained in a single package. This device changes alternating current (AC) to direct current (DC).
Reference Beam	In holography, the beam that exposes the film or plate. Usually a single beam of light.
Reflection Hologram	A hologram in which the object is behind the film/plate and exposed with a single beam that acts as both the reference and object beams.
Regulator	Any device/system that holds the circuit voltage to a specific level. May be either positive or negative.
Relay	An inductive based device that either opens or closes switch contacts when current is passed through its coil.
Resistor	Probably the most commonly used electronic component, it prevents (resists) the flow of electricity. Resistors are important in balancing power consumption.
Resonance Cavity	With "rod" lasers, this is a structure that is hollow and surrounds the rod and flash lamp assembly. The inside is usually highly reflective to concentrate as much light as possible on the rod.

SSS

Schawlow, Arthur L.	A Bell Lab researcher who worked with Charles Townes on extending stimulated emission to shorter wavelengths like light.
SDI	(Strategic Defense Initiative) Weapons program of President Ronald Reagan's administration designed to "knock down" incoming nuclear missiles with high-power laser beams.
Semiconductor	Any one of many electronic devices that is made from silicon substrate. These include diodes, transistor and, of course, integrated circuits.
Silicon Control Rectifier	(SCR) A semiconductor device, similar to the transistor, that latches after activation. It then takes a second signal to release the SCR. Useful for applications where you want the device to stay activated.
Single Hetero-Junction	The second architecture used for semiconductor lasers. Better than the first, it still was plagued by operational difficulties.
SMT	(Surface Mount Technology) Refers to the tiny sized components soldered to the component side of a circuit board that are being seen ever increasingly.
Snitzer, Elias	Credited with developing the first neodymium glass laser.
Solar Cell	Another name for a solar battery. These devices, of either selenium or silicon, generate electricity when light strikes them.
Solid-State	Designates any laser whose active medium is made of "solid" materials. All except the semiconductor lasers.

Sorokin, Peter P.	One of two IBM scientists that developed a special solid-state laser of calcium fluoride doped with uranium.
Spatial Coherency	A property of laser light that keeps the individual "cycles" locked together. That is, the cycle "crests" and "troughs" stay in step.
SPDT	(Single Pole—Double Throw) Type of switch that sends a single signal in one of two directions.
Speaker	An inductively-based electronic device that turns electric impulses into sound by vibrating a fiber or metallic cone.
Speckling	A property of laser light that gives the appearance of speckles of light around the "red dot." This is due largely to the very high resolution laser light reflecting off uneven surfaces.
Spectroscopy	The study of light spectra as produced by a spectroscope or other device.
Spillover	Applies to light in a holographic setup that falls outside the target (film or object). This is wasted light and should be kept to a minimum.
Split-Beam Hologram	A holographic configuration that splits the initial laser beam into two or more beams. This allows for a separate reference and object beam.
Split Power Supply	A power source that provides both positive and negative voltages. Useful with analog to digital converters and operational amplifiers.
Spontaneous Emission	The natural form of photon of radiation production through orbital jumping. It rarely if ever occurs in a large enough quantity to result in a population inversion.

SPST	(Single Pole—Single Throw) Type of switch that throws one signal in one direction. The oh-so-common "on-off" switch is one such device.
Sputtering	A flickering of the laser beam usually due to a deficiency of current, voltage or a "gassing out" situation.
Starter Voltage	With gas lasers, the voltage necessary to get the tube going. Normally, it will be substantially higher than the "running" voltage, and is also known as the "kicker voltage."
Stevenson, M.J.	One of two IBM scientists involved in the development of a uranium-doped calcium fluoride solid-state laser.
Stimulated Emission of Radiation	Orbital jump photon production by means of an artificial influence such as electricity or light. This is, of course, the "SER" in the acronym "LASER."
Stop Bath	A chemical in holographic processing used to halt the film development. Used before the bleach or fixer stage.
Synthetic Sapphire	Artificially grown crystals that have the same chemical composition as natural sapphire (corundum). Used for the rods in solid-state lasers.
Synthetic Ruby	Artificially grown crystals that have the composition of red corundum, or "rubies." Used for rods in ruby pulse lasers.

TTT

Telephone	I was unable to locate any suitable definition, and/or description of this device. So, I'll have to leave this one up to you. Nah, just kidding! You know what a telephone is.
Temporal Coherency	A property of laser light where the beam is emitted in very precise intervals. This, combined with "spatial coherency" helps keep the beam collimated.

Threshold	The point at which an operation is induced to occur. This is usually the result of an external voltage reaching a predetermined value.
Townes, Charles H.	Columbia University scientist and researcher who contributed heavily to early maser and laser development.
Transformer	An inductive device that transfers energy from one coil to another. Power and matching are two varieties of this component.
Transmission Hologram	A holographic configuration where the object is in front and to one side of the film/plate. These often use a single beam for both reference and object beams.
Transistor	A semiconductor junction that can provide both switching and amplification functions. The primary component of integrated circuits and the "solid-state" replacement of the vacuum tube.
Transmitter	A circuit that sends a carrier signal and/or information via radio waves, light or cable. In this context, the laser modulator and optics make up the light wave transmitter.
Tune	Refers to changing the output wavelength of a laser. Seen most often with dye liquid lasers.

UUU

Uitert, L.G.	One of three scientists credited with developing the first YAG laser.
Ultra-High Frequency	(UHF) I'll probably get an argument here, but this is the radio frequency spectrum between ABOUT 400 megahertz to ABOUT 800 megahertz. If you have a different spectrum, so be it.
Ultraviolet	Light extending off the violet end of the visible light spectrum.

Undulator	Also called a "wiggler," this is a magnetic field that causes a stream of electrons to oscillate back and forth. This results in the electrons emitting photons and is the principle behind the "free-electron" laser.
UPC	(Universal Product Code) The familiar "bar" codes found on most products. Theses are read with a laser to provide information to sales terminals and computers.

VVV

Very High Frequency	(VHF) The radio frequency spectrum from about 50 megahertz to about 400 megahertz. As with UHF, I expect some disagreement.
Vibronic	Applies to tunable solid-state lasers and refers to internal atomic transitions "tuning" the output wavelength. These lasers may some day replace dye units.
Volt	Named after Italian physicist Count Alessandro Volta, this is the basic unit of electromotive force.
Voltage Divider	A resistor network that divides an input voltage by a ratio determined from the resistor values to a lower voltage.
Voltage Doubler	A capacitor/diode arrangement that doubles, and rectifies, an input alternating current (AC) voltage.

WWW

Watt	Named after Scottish scientist James Watt, this is the "brute" behind the voltage. It actually started as a formula to define "power" as 1 amp flowing through 1 volt.
Wavelength	A term associated with the laser field that identifies the light's color. Usually measured in nanometers, micrometers can also be employed.

Wetting Agent — A chemical used in hologram processing that helps eliminate water spotting and other emulsion problems. Normally, this is the last step before drying.

Wiggler — Also called an undulator, this is the magnetic field that causes the electron stream in a "free-electron" laser to oscillate back and forth, causing electrons to emit photons.

XXX

X-Ray Laser — An extremely high-power laser that utilizes vaporized metal strips as the medium. These devices require another source of extreme power as the pump and have been part of several weapons programs.

Xenon — An inert gas element used for both lasers and flash tubes that pump lasers.

YYY

YAG — (Yttrium Aluminum Garnet) One of the most common of the solid-state laser "host" materials.

YLF — (Yttrium Lithium Fluoride) Another well respected solid-state "host" material.

ZZZ

Zeigler, Herbert — One of two scientists who helped Charles Townes develop the maser.

Zener Diode — A specialized diode designed to regulate voltage levels. This is done by "reverse biasing" the diode. However, that requires special diode construction, as most diodes will "burn up" under this condition.

Index

SYMBOLS

A

B

C

D

E

F

G

H

I

J

K

L

M

N

O

P

Q

R

S

T

U

V

W

X

Y

Electronic Projects for the 21st Century

by John Iovine

If you are an electronics hobbyist with an interest in science, or are fascinated by the technologies of the future, you'll find *Electronic Projects for the 21st Century* a welcome addition to your electronics library. It's filled with nearly two dozen fun and useful electronics projects designed to let you use and experiment with the latest innovations in science and technology. This book contains the expert, hands-on guidance and detailed instructions you need to perform experiments that involve genetics, lasers, holography, Kirlian photography, alternative energy sources and more. You will obtain all the information necessary to create the following: biofeedback/lie detector device, ELF monitor, Geiger counter, MHD generator, expansion cloud chamber, air pollution monitor, laser power supply for holography, pinhole camera, synthetic fuel from coal, and much more.

Projects
256 pages • paperback • 7-3/8 x 9-1/4"
ISBN: 0-7906-1103-1 • Sams: 61103
$24.95

RadioScience Observing
Volume 1

by Joseph Carr

Among the hottest topics right now are those related to radio: radio astronomy, amateur radio, propagation studies, whistler and spheric hunting, searching for solar flares using Very Low Frequency (VLF) radio and related subjects. Author Joseph Carr lists all of these under the term "radioscience observing" — a term he has coined to cover the entire field.

In this book you will find chapters on all of these topics and more. The main focus of the book is for the amateur scientist who has a special interest in radio. It is also designed to appeal to amateur radio enthusiasts, shortwave listeners, scanner band receiver owners and other radio hobbyists.

RadioScience Observing also comes with a CD-ROM containing numerous examples of radio frequencies so you can learn to identify them. It also contains detailed information about the sun, planets and other planetary bodies.

Communications Technology
336 pages • paperback • 7-3/8 x 9-1/4"
ISBN: 0-7906-1127-9 • Sams: 61127
$34.95

CALL 1-800-428-7267 TODAY FOR THE NAME OF YOUR NEAREST PROMPT PUBLICATIONS DISTRIBUTOR

Prices subject to change.